P9-CEH-000

ADDITIONAL PRAISE FOR

THE TWO-SECOND ADVANTAGE

• •

"The challenge of today's digital world isn't gathering data but making sense of it quickly. *The Two-Second Advantage* artfully explores how having the right information, in context and at the right time, can place you ahead of the game."

—David Stern, NBA commissioner

"Anyone interested in understanding the common thread of almost all long-term success should read *The Two-Second Advantage*. The authors capture your imagination with this well-written and lively exploration on how by just having unique insight prior to an event helps organizations make innovative decisions and keep their competitive edge."

—Chad Hurley, cofounder of YouTube

"In an environment where the velocity of change is faster than at any other time in history, a company's ability to capture *The Two-Second Advantage* can mean the difference between success and failure. Vivek articulates how leaders and organizations can use predictive processes to anticipate change and gain a competitive advantage that shapes the future of work."

—Francisco D'Souza, president and CEO of Cognizant

"An elegant exploration of how a company could in effect not guess, but anticipate what's about to happen in two months from now or even an instant with right information at precisely the right time. *The Two-Second Advantage* is one of those rare books that shapes our thinking about how companies and organizations should use technology to operate more like 'talented' humans."

—N. Chandrasekaran, CEO of Tata Consultancy Services

"What does the unique scoring ability of hockey great Wayne Gretzky have to do with leading the modern organization in the digital age—how valuable is a consistent competitive advantage driven by predictive power? With these engaging and insightful examples Ranadivé and Maney explore and explain how the leaders of 'Enterprise 3.0' are achieving sustainable competitive advantage through the use of predictive computing."

—Thomas H. Glocer, CEO of Thomson Reuters

"Critically important for today's business leaders. Customers are engaging with companies through an exploding number of channels, from mobile devices to the social universe. The concept that we can not only understand all that customer data, but make accurate and business-shaping predictions from it, puts this on the must-read list."

—Shantanu Narayen, president and CEO of Adobe Systems, Incorporated

"*The Two-Second Advantage* is a deft compilation of research and practical examples on how by having a little bit of the right information, at the right time and context, just far enough ahead is the key ingredient for success—in business and in other fields of human endeavor . . . the authors offer a vital perspective on how the available predictive capabilities can help make the world a better place."

—Klaus Schwab, founder and executive chairman of the World Economic Forum

THE TWO-SECOND ADVANTAGE

How We Succeed by Anticipating the Future—Just Enough

Vivek Ranadivé

Kevin Maney

CROWN BUSINESS
NEW YORK

Published in the United States by Crown Business,
an imprint of the Crown Publishing Group,
a division of Random House, Inc., New York.
www.crownpublishing.com

Library of Congress Cataloging-in-Publication Data
Ranadivé, Vivek.
The two-second advantage : how we succeed by anticipating the future just enough / Vivek Ranadivé and Kevin Maney.—1st ed.
p. cm.
1. Expectation (Philosophy). 2. Brain. 3. Human behavior. 4. Competition. I. Maney, Kevin, 1960– II. Title.
B105.E87.R36 2011
155.2'4—dc22 2011011497

ISBN 978-0-307-88765-8
eISBN 978-0-307-88767-2

Printed in the United States of America

JACKET DESIGN BY KEVIN TAM

10 9 8 7 6 5 4 3 2 1

First Edition

This book is dedicated to the thousands of TIBCO employees whose technical skills and business insights have helped form the founding principles behind the concept of the two-second advantage. They inspire me daily with their creativity, intelligence, and tenacity.

—Vivek Ranadivé

To my late father, Francis Maney Jr., who after forty-plus years still seems to hang around and help.

—Kevin Maney

Contents

PART II
TALENTED SYSTEMS

PART III
THE TWO-SECOND ADVANTAGE

Introduction

When the universe puts together a series of fortunate circumstances, it's probably a good idea to pay attention. Just such a set of circumstances led to this book.

Kevin Maney grew up in upstate New York and has played hockey since he was young. (He still plays and has all his own teeth.) He's almost exactly the same age and size as Wayne Gretzky. Somehow or other, Gretzky became the greatest player in hockey history, and Kevin . . . didn't. But for years now, Kevin has been fascinated with Gretzky's most pronounced talent: his ability to know what was going to happen on the ice a second or two before anyone else. In the mid-2000s, Palm cofounder Jeff Hawkins introduced Kevin to theories about how the brain works as a predictive machine, which led Kevin to explore the source of Wayne Gretzky's success in hockey. He'd begun sketching out ideas that had to do with predictiveness, talent, and Gretzky's brain for a new book he was contemplating.

Around the same time, Vivek Ranadive, the CEO of TIBCO Software Inc., noticed the arrival of a next iteration of technology that combined real-time computing, context, in-memory software, complex event processing, and analytics. Information about events happening in the moment could be correlated with historical data using software to predict future patterns. The result? New capabilities that could anticipate what was about to take place and act with precision before that moment arrives. TIBCO makes software technology that can do this stuff. Around TIBCO, Vivek started talking about his ideas. He felt that putting them into a book would help his employees understand his thinking and get technology and business people talking about the immense possibilities of this new capability.

Vivek and Kevin met in late 2009 in a TIBCO conference room, where Vivek told Kevin what he was seeing in terms of advances in technology. Kevin told Vivek about predictiveness and the human brain. Vivek realized that Kevin's ideas about the abilities of Wayne Gretzky's brain sounded a lot like the systems he believed companies had to implement to be competitive in the twenty-first century. And Kevin was intrigued that computer scientists were on a path to help explain talents such as those possessed by Gretzky.

They realized they had come across the right idea at the right time. Technology is reaching a breaking point, with too much data overwhelming computing's capabilities, and a new model of information technology is needed. The predictive nature of the brain is an expanding area of scientific discovery. And the intersection of the two—computer science and neuroscience—is an increasingly hot field that's likely to give birth to the next generation of information technology. On their own, neither Vivek nor Kevin would've seen the connections between technology and the brain so clearly. Together, they found the fields created a perfect synergy for the 2010s. This book is the result of their collaboration.

The authors would like to acknowledge some individuals for their valuable help on this project.

Vivek Ranadivé: With any project, there are many individuals to thank behind the scenes and this book is no different. The support of my family and friends and their help and feedback in transforming my thoughts into a concept worth publishing has been invaluable. Roger Scholl, my editor, and Kevin Maney, my coauthor, have been incredibly patient and insightful, and have not only done the hard work of tying all the threads together, but have made significant contributions to the ideas in this book. I would also like to thank several colleagues who contributed greatly to this book's content: Don Adams, Kal Krishnan, Matt Langdon, Ram Menon, Matt Quinn, Murat Sonmez, Raj Verma, Srini Vinnakota, and many others too numerous to mention. My gratitude to Kevin Tam and Anthony Zambataro for their outstanding cover design, and to Jennifer Quichocho, my assistant, for all her help.

Kevin Maney: I would like to thank Dan Fost and Russ Mitchell for their research work. Jeff Hawkins gets a special nod for introducing me to the idea of the predictive nature of brains in the early 2000s, and for putting up with my asking him more about it over and over again through the years. Thanks, too, to neuroscientists Jim Olds, Stephen Grossberg, and Paula Tallal for their general advice and input. Thanks to Roger Scholl, the editor on this book (and my previous book, *Trade-Off*), for having faith in the project and seeing it through. Sandy Dijkstra, our agent, helped make it happen. Thanks to the folks at TIBCO who from time to time set aside running a fast-growing company to help the book along, including Holly Gilthorpe, Ram Menon, Don Adams, and Srini Vinnakota. Finally, much appreciation to Kristin, Alison, and Sam for understanding when book writing had to take over weekday evenings and Sunday afternoons.

PART I
TALENTED BRAINS

WAYNE GRETZKY'S BRAIN IN A BOX

In the 1981–82 hockey season, Wayne Gretzky broke the National Hockey League record by putting ninety-two pucks in the net. At the time, he stood five feet eleven inches tall and weighed 170 pounds—a wisp compared to the average NHL player. "I look more like the guy who bags your groceries at the local supermarket," he said about himself.[1] He wasn't even a particularly great athlete, in a purely physical sense. "Our team doctors tested my endurance, strength, reflexes and flexibility with machines, bicycles and drills," Gretzky told an interviewer. "They tested every guy on the team and I did BAD in all the tests."[2] Yet Gretzky holds just about every hockey scoring record there is. He is the best player in the sport's history.

And here's the crazy thing: Gretzky didn't get so good in spite of his unimpressive physical attributes—he became so good *because* of them.

Gretzky grew up in Brantford, Ontario, and started skating on the nearby river when he was two. He played hockey with the local kids every chance he could in winter—an hour before school, a couple of hours after school, another hour or two after dinner. His father, Walter, taught him and coached him, although he didn't push him. "I practiced all day because I loved it," Gretzky said. When he was six, he tried out for the Brantford Atom League for ten-year-olds and made it. One photo from that season shows Gretzky skating with his teammates, his head about as high as the numbers on most of his teammates' jerseys.

That first year, Gretzky scored one goal. The next year, he scored 27 goals; the next year, as an eight-year-old, 104; then 196; and when he was ten, Gretzky scored 378 goals in sixty-nine games. Yet he was always one of the smallest, scrawniest players in the league. No one had seen anything like it. Newspapers and magazines rushed reporters to Brantford to write stories about him.

Gretzky quickly moved to higher-level leagues with much older, beefier guys. Since Gretzky couldn't unleash physical prowess on these opponents, he developed a different kind of weapon: his brain. "When I was five and playing against 11-year-olds, who were bigger, stronger, faster, I just had to figure out a way to play with them," Gretzky explained. "When I was 14, I played against 20-year-olds, and when I was 17, I played with men. Basically, I had to play the same style all the way through. I couldn't beat people with my strength; I don't have a hard shot; I'm not the quickest skater in the league. My eyes and my mind have to do most of the work."

He added: "I had to be ahead of everybody else or I wouldn't have survived."[3]

Gretzky's father taught him anticipation, and Gretzky memorized hundreds of tricks and shortcuts—and then perfected them, because he had no other way of succeeding on the ice. The more he played, the more that sense of anticipation became instinct.

Before long, he could hold the whole evolving situation—everything that was happening on the ice and the movement of every player—in his mind. "When you're 170 pounds playing with 210-pound guys, you learn to find out where everybody is on the ice at all times," Gretzky noted. Being small forced Gretzky to develop an exquisite hockey brain. He built a predictive model of hockey in his head, so that as a game unfolded he could use memories of past games and tactics, and a reading of the immediate situation, to predict what would happen next.

Every other good player does this to some extent. But Gretzky could do it just a little bit faster and a little more accurately than everyone else. Lots of kids growing up in Canada had skated for hours from the time they were preschoolers. Lots of kids had fathers who coached and drove them. Most of those kids had bigger, stronger bodies than Gretzky. Yet none of them became a Wayne Gretzky, because none of them developed the predictive brain Gretzky did.

He truly was able to understand what was going to happen an instant or two before anyone else on the ice—and skate to where the puck was going to be. That was his famous line: He'd say that he doesn't skate to where the puck is—he skates to where it's going to be. Commentators would often say that Gretzky seemed to be two seconds ahead of everyone else. That capability drove Gretzky's phenomenal success in the NHL. Gretzky went on to lead the Edmonton Oilers to four Stanley Cup championships.

"He reads where other people are going to be," said Grant

Fuhr, who played with Gretzky on the Oilers. "People don't even think of a play, because they don't think that play is possible. And Gretzky makes that play. He'll pass to a place, not a player. Somebody'll be heading to a place, and Wayne knows they can score from that spot, and that's where (the puck) goes."[4]

But what's going on inside Gretzky's head from a *scientific* point of view? Are there lessons from Gretzky that have implications for, say, running a department store?

Like Gretzky on ice, the most successful people in various fields make continual, accurate predictions just a little ahead of and a little better than everyone else. It is the one common denominator of almost all consistent success. Talented people don't need to have a vision of the future ten years out or even ten days out. They need a highly probable prediction just far enough ahead to see an opening or opportunity an instant before the competition. That's true for athletes, artists, businesspeople, or anyone in any field.

Metaphorically, the prediction only needs to be two seconds out—though the actual time may be hundredths of a second or several minutes, depending on the situation. In other words, talented people have a two-second advantage. (In his 2005 best seller *Blink: The Power of Thinking Without Thinking*, author Malcolm Gladwell describes how judgments made in two seconds are often more accurate than those made after months of analysis. "It's a system in which our brain reaches conclusions without immediately telling us that it's reaching conclusions," Gladwell writes.[5])

This concept maps to theories in neuroscience about intelligence that have solidified over the past two decades at research centers such as the Redwood Center for Theoretical Neuroscience at the University of California at Berkeley; in the work of neuroscientists such as Stephen Grossberg at Boston University and Paula Tallal at Rutgers University; in experimental projects funded by the Defense Advanced Research Projects Agency (DARPA); and in a number of other research centers.

First, these scientists have found, the brain forms memories assembled from experiences. Those experiences get stored as patterns and assembled into quickly accessed chunks of information. The more experiences are repeated, the stronger and more complex the patterns become. When Gretzky, for instance, saw an opposing goalie move a certain way, the image fired up an instantaneous, complex pattern built out of everything Gretzky had experienced and stored in his memory.

These capabilities don't hold true just for superstars in sports. In everyday existence, our senses constantly send information to our brains. The brain uses this stream of information to fire up stored patterns of memories, telling us, This looks familiar—here's what's probably going to happen next. The brain tests what it thought would happen against what actually happens, and adjusts—and makes new predictions. Brains can perform this sequence in milliseconds, constantly.

When walking up stairs, your brain recognizes from previous patterns that the next stair will be as high as the previous stair, so it directs your foot to follow that prediction. That prediction is what allows you to walk up a set of stairs in the dark without thinking much about it. But if one stair is a little higher than the others, you may trip and have to start paying closer attention.

The human brain is a predictive machine. Intelligence is prediction. This is a relatively new concept in neuroscience, coalescing into broad acceptance only in the 1990s and 2000s. While the connection between prediction and general intelligence is generally understood, an even newer—and largely unexplored—idea has emerged in neuroscience: exceptional predictive capability is what drives talent. Up to now, only a few empirical studies have focused on that link, but a number of scientists and psychologists have suggested theories about talent and predictiveness. Anecdotally, when talented people are asked about their abilities, they often describe a superpredictive capability—as Gretzky has in the past.

Most successful people are really good at making very accurate predictions—usually about some particular activity—just a little faster and better than everyone else. We see it all around us. The salesman who sells more than anybody else has developed a talent for anticipating people's reactions to his pitches, allowing him to steer the conversation before it gets off course. The teacher who seems to get the most out of her students with the least effort has developed predictive models in her head for how kids behave and respond to certain teaching methods.

The great jazz saxophonist Joe Lovano talked to us about hearing notes before they're played. A man named Mystery—self-described as the "world's greatest pickup artist"—said he can anticipate how women are going to react in a bar. Tom Menino, the seven-term mayor of Boston, told us about the mental model of Boston he's built in his brain, allowing him to instantly anticipate how a proposed law or building will affect the city.

There are two ways to acquire this kind of prediction-based talent.

One is to inherit it. Some people have brains that are wired to do something better than the rest of us. Their neurons fire and connect faster and organize themselves more efficiently for a particular task. They can master a skill without much training—like the kid who gets straight *A*s without toiling over homework or wins piano competitions despite barely practicing. At the extreme end of inherited talent are savants, and such talent is rare and comes with difficulties.

The other way to gain a more predictive brain is to develop one. Perhaps a person doesn't have as much raw talent as some. But through thousands of hours of practice and hard work and field testing, that person can craft an efficient, effective predictive model in his or her field. Such people practice until they get it right. They become many of the world's success stories—the writers who penned ten failed novels before writing that magical book that took off; the entrepreneur who failed in several businesses before founding a hot start-up.

But to achieve the success of a Gretzky, most people need both the natural wiring and the hard work, forging them together through a singular combination of circumstances.

"Gretzky was uniquely lucky," George Mason University neuroscience professor James Olds told us. Gretzky's brain had the advantage of great circuitry, to be sure. Because of his love of playing hockey, plus the long Canadian winters and his father's coaching, Gretzky got thousands of hours of practice that seared the dynamics of the sport into his memory. And then his small size gave him a reason to rely on his brain instead of his athleticism, stoking him to build better and better predictive models until he could out-anticipate anyone in the game.

This, ironically, allowed Gretzky to not have to think so much during a game, allowing his brain to respond faster. Through his career, he continued to master ever greater complexity about the game, eventually storing whole symphonies of complex movements in single chunks. Think of it as being a little like baking. Other players, when taking in a developing situation on the ice, essentially were seeing the flour, the eggs, and all the other individual ingredients and had to think about how to put them all together to make a cake. Gretzky just saw a cake.

Instead of constantly having to access all the information he had stored about hockey in his brain during games—which would've taken too much computational time and effort—Gretzky was able to access whole chunks of information that he'd already assembled, analyzed, and understood. The informational chunks formed a well-honed, efficient mental model of hockey. All Gretzky had to do during a game was reference his mental model and then let the information unfolding in the game flow in through his senses. When he recognized the way a play was going, it would activate chunks of knowledge based on what he'd seen before. That would in turn generate a prediction: this is what's probably going to happen. His brain would test what he was seeing against the prediction, perhaps taking into account that one of his teammates seemed a little slower that game, or the unexpected positioning of a rookie defenseman. Then, in a flash, the predictive model in his brain would adjust to the action on the ice and he would know, with astounding accuracy, what was going to happen next.

Gretzky didn't try to create a plan for the whole sixty-minute game, or even think about what he was going to set up for thirty seconds from now. All he had to do was predict what was about to happen in the next instant and do it a little bit faster and a little more accurately than anyone else. It gave Gretzky a gigantic advantage. It was his two-second advantage.

Imagine if a company could, in effect, skate to where the puck is going to be—not guess where it might be three months from now but correctly anticipate what's about to happen in an instant. Some entities are already moving in that direction. Sam's Club does an amazing job of knowing what its members are going to want to buy when they walk into the store. The East Orange, New Jersey, police department is getting better at predicting where and when a crime is about to take place, so it can have a police car drive by, preventing the crime from ever occurring. Such examples are not at the sophisticated level of Gretzky's brain during a hockey game, but they are an important leap in that direction.

Computer experts realize that the way the two-second advantage worked in Gretzky's head holds the key to how it can work in technology and ultimately change how companies and other organizations operate.

On March 28, 1955, *Time* magazine reported on a new generation of machinery called computers. The cover featured a drawing of IBM's Thomas Watson Jr. in front of a cartoonish robot over a headline that read, "Clink. Clank. Think." The story marveled at a computer built by IBM working inside a Monsanto office building. "To IBM, it was the Model 702 Electronic Data Processing Machine," the story reported. "To Monsanto and awed visitors, it was simply 'the giant brain.' "[6]

Technologists have long tried to build computers that can do brainlike things. They've worked on artificial intelligence and robots and on making computers that can beat grand masters at chess. But those projects have all had narrow success at best. The basic structure of computers works very differently from that of brains. Computers can do some things better than humans, like instantly calculate long equations or sort through millions of documents looking for a few key words. But they can't do some of the simplest things even a three-year-old can do—like knowing that a line drawing of a cow and a real cow are both a cow. Computers definitely can't match the brain's higher-level processes, like putting disparate ideas together in a flash of inspiration. Building a computer that thinks like a person is a long way out—and perhaps a quixotic quest in the first place.

And yet computer scientists are learning from human brain research and are building computer systems to operate in new ways borrowed from the human predictive model. These systems, in their own way, build memory chunks and generate behavior based on predictions. Sensors can feed information back to the computers to both build patterns and test predictions.

Forward-thinking companies are starting to use these new systems to operate more like talented humans than bureaucratic organizations. These companies can use technology to sense what's happening in the market, constantly adjust, and act just a little bit ahead of time—a two-second advantage.

As it turns out, getting just a little bit of the right information just ahead of when it's needed is a lot more valuable than all the information in the world a month or a day later. Using a database to analyze piles of data after the fact would be like Wayne Gretzky scouring through all his hockey memories to analyze why he didn't score in the last game and make a plan for the next game. While that might be valuable, it's not enough anymore. Enterprises will want to anticipate like Gretzky, using an efficient "mental model" to get a little ahead of events and make instant judgments about what to do next. Companies will be able to anticipate customers' needs. Stores will no longer carry too much or too little of a product. Law enforcement will be able to stop criminal acts before they happen.

A number of trends are coming together to facilitate two-second-advantage technology.

For fifty years, we've lived in a world of database technology. Corporations and government agencies collect information from individual interactions (forms filled out, reservations made), transactions (at ATMs, on the Web, through credit card purchases), and recorded events (baseball scores, hurricane readings from the Gulf of Mexico, airline departures from LAX). The data gets fed into a structured database, which can mix, match, and analyze it to make discoveries about things that happened.

A database might tell a retailer that it sells 50 percent more Pampers in August, suggesting the store should stock up that month. Or a database could tell an airline that when it lowers prices by twenty dollars on a particular route, it steals a chunk of market share from its competitors. A census is a monstrous database that can identify patterns in a nation's population every ten years.

Databases can help an executive make informed decisions about what to do next based on past outcomes, and that's valuable. And databases have gotten increasingly real time. A couple of decades ago, an executive had to put in a request for information from a database and wait a day or a week to get the results. In 2011 databases can update information on the fly and instantly answer a query from an executive with a flood of information about what happened earlier that day.

Database technology is critical to the operation of nearly every enterprise of any size everywhere on the planet. Yet databases have major handicaps in today's world. They're inherently focused on the past. They analyze what's already happened, not predict what's about to happen. And databases are about to get overwhelmed by crushing waves of data from a constantly expanding number of sources. Database technology won't be able to keep up.

In 2010 more than 1,200 exabytes of digital information was created. A single exabyte is equal to about one trillion books. Every two years, the volume of data created quadruples. About 70 percent of it is created by individuals, including profile information on social networks, videos on YouTube, tweets on Twitter, music preferences on Pandora, and location check-ins on Foursquare. The rest is coming from an ever-expanding universe of sensors. These include chips placed on buoys to keep tabs on a bay's water, RFID tags on luggage that tell an airline every time a bag is loaded or unloaded, and the billions of cell phones in the world—each of which constantly tells the cell company where people are and how they move around.

At the same time, storage technology is improving so fast, it will be possible to gather and store all of this data that comes roaring in. While more data can certainly be valuable, too much can get overwhelming. If database technology has to sort through all the data to answer every query, it will get bogged down. Answers will come too slowly. Just as Gretzky can't search every memory during a game, a business can't search all its data each time it needs an answer.

Rapidly escalating stores of data might be less of a problem if computer processing power could increase fast enough to keep up. But that's not likely. Since the 1970s, processing power has improved at a pace described by Moore's law: roughly twice as many transistors can be packed onto a microprocessor every eighteen months. From the 1980s to the late 2000s, computer systems got hundreds of times faster. But the individual transistors have now gotten so small—less than a dozen atoms across—that they can't get much smaller. The kind of technology that currently runs almost all computers can only get two or three times faster.

To handle the coming data onslaught, technologists are pursuing alternative kinds of computing. The most promising path is to develop more brainlike computers—technology that will use data to build models as Gretzky did, learning from data but not relying on the entire database. The technology will be able to read real-time events and predict what will happen next.

That kind of speed and agility in technology is increasingly critical. In business, government, and everyday life, time to react is dropping. Competition is driving the world to work at an ever faster pace. No one can afford to react too late based on information that's too old. The new competitive advantage will be an ability to anticipate events based on information about what's happening right now.

Before the Internet, we were in an era we call Enterprise 1.0. In a bank, for instance, customers would come in all day long—no ATMs!—and tellers would pile up pieces of paper tallying transactions. At the end of the day, the branch manager would account for everything, then send the information to headquarters, where information from all the branches would be assembled and calculated. Getting a report on the state of affairs at the bank might take days or weeks. Reaction time to any single event could be measured with a calendar.

Computers and the Internet ushered in Enterprise 2.0. Every transaction became a bit of digital data. As computers and networks got more powerful, that data could be calculated and analyzed faster and faster, to the point where the bank CEO could look at a computer screen and see the money flowing in and out of his banking company in almost real time. Reaction time to any single event could be measured with a stopwatch.

We're entering Enterprise 3.0. Now every *event* can become a bit of digital data. A transaction is one kind of event, but there are many others too. Every time a customer logs on to the bank's Web site, even if no transaction is completed, that's an event. Cell phone signal analysis may tell a bank how many people walk by a branch every day—more events. Debit card purchases at far-flung retailers are events. A bank should be able to recognize patterns of events and anticipate what a customer might want next, proactively capturing that business. Reaction time to any single event will have to be measured with a time machine—because the idea is to act in anticipation.

In the era of Enterprise 3.0, making decisions based on information even just a few seconds old could be disastrous. Trying to make decisions based on *all* the events coming in would be mind-bogglingly difficult. The new systems need the right information in the right place at the right time, so they can anticipate what's coming next.

The basic idea of using data to be predictive in business doesn't come out of thin air. Companies have been deploying mathematics and software to try to foresee events for decades. Statistical analysis proved that events could be predicted within levels of probability—like the mean time before failure of mechanical equipment or the likelihood of people within a given zip code to respond to a certain direct-mail campaign. Big software systems in categories such as business process management (BPM) and customer relationship management (CRM) have tried to gather all of what's going on in a corporation and help managers understand when, for instance, an assembly line will need a new shipment of parts or a customer might be ready to buy an upgraded product.

In more recent years, companies have employed analytics to uncover trends from historical data. Analytics can look at a person's pattern of past spending and bill paying, compare it with patterns of millions of other consumers, and make a statistical prediction about whether that person will default on a loan. Analytics help airlines predict demand so they can adjust schedules to make sure planes fly as close to 100 percent full as possible.

We're not suggesting that we're inventing the idea of predictive technology. We're not saying anyone should throw out their BPM, CRM, or analytics systems. There will always be great value in making projections that are days, months, or years out—just as people need to make long-range plans or coaches need to make game plans they think will work against an upcoming opponent.

But in today's world, enterprises need something more. They need that instantaneous, proactive, predictive capability of a Gretzky. In the 24-7 ongoing rush of events, enterprises need to be able to put their mountains of data to the side and act using small, efficient "mental" models that can spot a series of events, anticipate what's about to happen, and initiate action in a split second.

It's not a pipe dream. A number of companies are implementing some of the first predictive systems, aimed at getting the right information to the right place just a little ahead of time.

Southwest Airlines is developing a system that will let it watch its inventory of planes, the weather, ticket prices, and other factors and constantly adjust—perhaps sensing that a storm is coming and refiguring the airline's entire schedule and moving passengers to different flights before routes snarl.

Xcel Energy is testing a system in Boulder, Colorado, that uses two-way meters and outage monitoring equipment to build memories of how electricity moves through the distribution grid and what seems to trigger problems. That way, the grid can react to an event it sees coming and reroute electricity or increase capacity just a little ahead of what otherwise would become a power outage.

DARPA, the Pentagon's futuristic research arm, is funding a program called SyNAPSE, which has a long-term goal of creating an entirely new technology architecture that can work like the brain.

"We are in a major revolution," Boston University's Grossberg told us. He is perhaps the best-known researcher crossing over between neuroscience and computer science. He has been studying brains for more than fifty years, and in an interview, as he scooped up sushi at a restaurant near his office, he talked like a fired-up youngster. He is advising two of the teams involved in the SyNAPSE program. He and his colleagues are churning out papers about how the brain turns information into thought and then action. "Building models of the brain into technology is not a future activity," Grossberg said. "It's current. We're building real-time systems. The problem is an always-changing world, and we need systems that can deal with unexpected environments."[7]

In other words, we need Gretzky's brain built into technology.

On Christmas Day 2009, Dutch filmmaker Jasper Schuringa relaxed in his seat on Northwest flight 253 from Amsterdam to

Detroit. The flight was making its final approach to Detroit's airport when Schuringa was startled by what sounded like a firecracker going off. "First, it was just a bang," Schuringa told CNN. "And you're trying to look around, like, where's this bang coming from?" Schuringa noticed a man on the left side of the aisle, sitting still while on fire. "A normal person would stand up, and he wasn't standing up," Schuringa said. "So then I knew, this guy is trying to do something."

The guy was Umar Farouk Abdulmutallab, a twenty-three-year-old Nigerian who had ties to radical terrorist group al Qaeda. Abdulmutallab was carrying PETN, or pentaerythritol tetranitrate—enough of it to blow a hole in the aircraft—sewn into his underwear. But the device he used had failed to detonate, instead setting off the fire. Schuringa jumped over the passenger next to him and lunged for Abdulmutallab, wrestling the device out of the Nigerian's hands. Crew members and other passengers jumped on Abdulmutallab, stripped him, and handcuffed him. These actions no doubt saved the lives of the three hundred people on board.

But then the question was, Why did Schuringa and his fellow fliers have to act so heroically? So many pieces of information should've led authorities to stop Abdulmutallab before he ever boarded that plane.

Four months before the attempted bombing, the National Security Agency (NSA) had intercepted phone conversations between al Qaeda leaders who were talking about using a Nigerian bomber in an attack. Around the same time, U.S. counterterrorism agents had learned that al Qaeda had figured out how to hide PETN in underwear. That November, Abdulmutallab's own father had gone to the U.S. embassy in Nigeria and told officials he was concerned that his son had come under the influence of militants and might do something rash. Meanwhile, British authorities had denied Abdulmutallab a visa because he had applied to attend a fake institution.

Yet despite investments in technology and plans to share information, all those clues—and many more—about Abdulmutallab remained in separate databases, which were unable to put the pieces together on their own. The National Counterterrorism Center employs specialists who can tap into more than eighty databases, but it's up to the specialists to conduct the searches and crunch the information to match up clues. In a 2008 report by the House Committee on Science and Technology, investigators found the system to be ineffective.

"The program not only can't connect the dots, it can't find the dots," Representative Brad Miller, a Democrat from North Carolina, said at the time.

The solution isn't to try to build a computer system powerful enough to constantly sort through the avalanches of data flowing in from every government agency and security outpost, always trying to match it with other data stored somewhere else. That would take too much time and too much processing power. Instead, federal agencies need a system that would work more like Wayne Gretzky's brain, constantly categorizing data and seeing relationships, and then using that to build "chunks" and an ever-evolving model of how things work in the terrorism universe. Streams of data from all over the world would come in, much the way a person takes in vast amounts of data through his or her senses, and the new data would constantly be sifted through the predictive model that the system constructed. Like Gretzky on ice, the system could then react in real time to what it was "seeing"—and conclude that if an al Qaeda operative was discussing a Nigerian bomber, and a Nigerian told a U.S. official about his dangerous son, and the British denied a visa to that man's son, then the son should not be allowed to board an airliner destined for a U.S. city.

The system wouldn't skate to where the puck is—it would skate to where the puck is going. That approach could prevent the next terrorist attack from ever getting started.

Ones, Twos, and Cortexes

In the winter of 1996, Ben Horowitz was toiling as an un-heralded product strategist at Netscape Communications when he opened a scathing e-mail from his boss, Marc Andreessen. Netscape's public offering just months earlier had ignited the dot-com craze, and as a Netscape cofounder, Andreessen had gotten rich and appeared on *Time*'s cover, sitting on a throne, feet bare—the very portrait of a cocky twenty-four-year-old tech wunderkind.

Horowitz, though, was not intimidated by his boss, and he had been irked to learn that Andreessen had leaked news to a trade publication about an upcoming software release Horowitz's team had been working on. So Horowitz had sent Andreessen a note that simply said, "I guess we're not going to wait until March 7"—the date of the planned announcement. Now a flaming e-mail came back from Andreessen: "We are getting killed

killed killed by Microsoft! You're destroying the value of the company and it's 100% server product management's fault. I'm just trying to help. Next time, do the fucking interview yourself. Fuck you. Marc."[8]

More than a dozen years later, Andreessen and Horowitz sat around a table in Manhattan on a muggy June day, sipping iced tea, laughing uproariously as they retold the story. "This is why I should not run a company," Andreessen said. The two became tight friends at Netscape. Andreessen cofounded his second company, Loudcloud, with Horowitz—and made Horowitz the CEO. In 2009 the pair launched a high-profile venture capital firm, Andreessen Horowitz. While Andreessen has the star power, even he will tell you that teaming with Horowitz to invest in companies was critical. Around Silicon Valley, Horowitz is considered a connoisseur of start-up CEOs. Hardly anyone can read them better, pick them better, or coach them better. He writes a closely watched blog that's mostly about business leaders and the decisions they have to make. Having been a successful CEO at Loudcloud and its successor company, Opsware, plus having invested in numerous start-ups such as Twitter, Horowitz seems to understand this breed to the core.

One thing he's come to believe, he told us, is that there are two types of people in the top ranks of companies:

- There are ones.
- And there are twos.

And, he believes, the brains of each type function entirely differently.[9]

Ones are predictive. Twos have to rely on mountains of data to figure out what they think. Ones should be CEOs, and twos should not.

As Loudcloud's CEO, Horowitz loaded up on knowledge about his company. Not data but knowledge. He didn't have to recall every sales statistic or financial number, but he wanted information about the products, the customers, the employees, the challenges, and so on to flow through his brain, which helped him feel wired into the company. Basically, he chunked together complex patterns, built a model of his company in his head, and governed it with his personal set of rules about values and how a CEO should make decisions. When a tough problem presented itself, Horowitz almost instantly knew what to do. "I'd have a feeling about it and immediately feel I was right," he told us. "It's a gut feeling based on a massive amount of knowledge." Horowitz might ask for more input or data, but it usually didn't change his mind. And if speed was necessary, he was confident of his instinct. He could predict what was going to happen and usually be right.

This, Horowitz believes, is the way a start-up tech CEO has to work. "When we invest in companies, I'm looking to see if the company has somebody who has that capability—that speed and quality of decision making," Horowitz said. "They need to be a one, and not a two."

In Horowitz's universe, ones tend to be founders. They are bullheaded and courageous. They tell people what they think, not what they think people want to hear. They see openings and get flashes of creativity—like Gretzky in a hockey game. They can take in everything that is happening at a company and see it from a higher level, the details blurring into instinct.

Make no mistake: Twos are still extremely important to a company. Ones need twos. The twos pay attention to details; they get things done. But, Horowitz says, the twos almost never develop the predictiveness of a one. "They tell me what I want to hear," Horowitz said. "They lack confidence in their intuition. They don't have the courage. Twos don't have that predictive capability. They want to have data to base a decision on. When does 'We *think* this is right' trump 'We *know* this is right'? For a two, it's never."

The twos, lacking that predictive model, want to constantly go back to the mountains of data, but accessing it and sorting through it takes time and often doesn't improve the decision. To Horowitz, a CEO who is a one has a two-second advantage; a two CEO does not.

"Every decision that a CEO makes is based on incomplete information," Horowitz explained in one of his blog posts. "In fact, at the time of the decision, the CEO will generally have less than 10% of the information typically present in the ensuing Harvard Business School case study. There is never enough time to gather all information needed to make a decision. The CEO must make hundreds of decisions big and small in the course of a typical week. The CEO cannot simply stop all other activities to gather comprehensive data and do exhaustive analysis to make that single decision."[10]

The secret to being a great tech CEO is having an efficient, agile mental model that can quickly predict what's going to happen and be right most of the time. Microsoft cofounder Bill Gates is a one; his successor, Steve Ballmer, is a two. Apple CEO Steve Jobs is a classic one. The CEOs who ran Apple—and nearly killed it—during Jobs's exile from the company were all twos.

Back in 1979, Jeff Hawkins was a newly minted engineer, just out of Cornell University. He wanted to work for Intel. But that fall he picked up an issue of *Scientific American* devoted to the brain, pored over the articles, and carefully read the concluding piece in the magazine by Francis Crick, the neuroscientist and molecular biologist who was one of the discoverers of DNA. Hawkins summed up Crick's article this way: "He said, this is all well and good, but we don't know squat about brains," Hawkins said.[11]

Hawkins took that as a challenge and started studying brain theory. He couldn't find a brain theory graduate program he liked, so he went into the computer industry instead. But he continued his brain research on the side, hoping to eventually steer himself back into brain science. In the intervening years, Hawkins made a fortune, inventing some of the most important mobile products in history, including the PalmPilot and Handspring Treo. Eventually, he dovetailed his twin interests in computers and brain science, convinced that a good theory of the brain would help in the quest to build intelligent machines. In 2002 he founded the Redwood Neuroscience Institute in Menlo Park, California (which became the Redwood Center for Theoretical Neuroscience in Berkeley in 2005), and started writing a book, *On Intelligence*, which was published in 2004.

In speeches and articles, he claimed that neuroscience had produced a ton of data about the brain's workings but no good theory about how intelligence worked. For decades, the belief among scientists had been that intelligence was defined by behavior. "And that's wrong," Hawkins said. "Intelligence is defined by prediction."[12]

On Intelligence drew on the most advanced thinking about brain science and laid out a theory of intelligence that turned out to be highly influential. Hawkins calls himself a "cortical chauvinist" because of the way the neocortex (or simply "cortex," in the common shorthand) plays such a central role in intelligence. "Prediction is not just one of the things your brain does," Hawkins wrote in the book. "It is the *primary function* of the neocortex, and the foundation of intelligence."[13] (The italics are his.)

Most of nature's creatures have a brain that oversees the body's central nervous system. At a moderately high level, like that of an alligator or squirrel, the brain guides a set of behaviors we'd call instinct. It tells the alligator how to hunt and the squirrel when to pack in nuts for the winter. More social animals, like chimps or dolphins, developed more complex behaviors, including ways to communicate with one another and compete for mates. Such social interaction drove the evolution of a higher-level brain that sat on top of the "old brain." At the same time, social beings that developed better higher-level brains led more successful lives and passed on their better-brain genes, boosting the evolutionary process. As humans took complex behavior to radical new levels, the human neocortex grew in sophistication and size. It became so big it had to start folding in on itself to fit inside the human skull, which is why a brain looks the way it does. The neocortex is the outermost layer of the brain, and the human neocortex is about twice the size of any other mammal's. It's where the really interesting stuff happens in the brain: language, attention, consciousness, and memory.

The human neocortex is made up of six thin layers of cells wrapped around the old brain. If you were to peel those layers off the old brain and press them out on a table, pulling all the wrinkles flat again, you'd end up with something roughly the size of a place mat. The neocortex is about four hundred square inches in area and one-sixth of an inch thick.[14]

The cells of the outermost, or bottom, layer of the neocortex handle sensory perception—things like visual signals coming from the eye via the optic nerve or sounds traveling along auditory nerve pathways. The bottom layer handles these signals by breaking them into very small, discrete elements. A group of cells in the cortex might be responsible for nothing more than registering a tiny vertical line segment in a certain component of your visual field. Imagine a sentry staring at the world through a straw. His sole task is to report when a specific geometric image shows up in the other end of the straw. That's it—one discrete processing job for each cluster of neurons.

When a sensory signal comes in, the bottom cells don't even analyze it. They simply pass it up to the next layer in the neocortex, where the brain begins to process and combine signals from other cells. That process gets repeated—cells in the second layer hand information up to the third, which combines information at yet a higher level, and gradually the brain starts to assemble discrete pieces of sensory information into an image or idea, which it compares against the millions of images and ideas in memory. This is how shapes become objects and objects become concepts. A line segment in a certain part of your field of vision gets combined with other signals from nearby cells—say, other arcing lines that connect to the original—and your brain eventually says, Ah, I recognize this. It's a circle.[15]

When that happens, the lower and upper cells "fire," or signal each other with a tiny electrical pulse, which neuroscientists can measure in devices like functional MRI (fMRI) machines. You can think of that firing as, literally, the spark of recognition. Researchers call it "resonance"—a kind of electronic agreement between cells at different layers.

In this way, the neocortex works a little like the division of labor at a corporation—the lower levels (like workers on a factory line) handle very specific jobs with a small universe of options. And the higher levels of the neocortex (like the top executives) operate at a more idea-driven, conceptual level. The top levels receive signals from multiple sources, aggregate them, compare them to what's happened in the past, and say, Okay, I've seen something like this before.

One more significant concept comes into play: "invariant representations." It essentially means that the incoming input doesn't have to be a perfect match. You can see a lot of circle-shaped objects (small ones and large ones, wedding rings and truck tires, hula hoops, piston rings, shoelace grommets, etc.), and your brain recognizes all of them as circles. You can even see part of a circle—say, a small segment of a bike wheel that's largely hidden behind a brick wall—and your brain automatically fills in the rest. It's as if the brain had a big folder stored in memory with the generic label "circle," and that folder could handle a lot of variation in the stuff that goes in it. This is also how your brain recognizes that a real cow and a line drawing of a cow both fall under the category of "cow," even though they don't look much alike.

This flexibility is one of the key differences between the way a brain works and the way a computer works, said Boston University's Grossberg—who was a significant influence on Hawkins. The brain can classify things in patterns without being locked into rigid rules about those patterns. As Grossberg put it, the brain is flexible enough to understand not merely rules but "rules plus exceptions."[16] In tech circles, this is called fuzzy logic. Computers, which operate on the cold calculus of rules and absolutes, are terrible at fuzzy logic. (Jeff Hawkins has a new company, Numenta, that's working to change that—a topic for later in the book.)

The top cells of the neocortex don't just patiently wait for signals to be handed up to them for identification. This is where predictiveness comes in. The top layers also send signals back down to the lower layers of the cortex. Each neuron has connections to thousands of other cells in the brain, and the top layers prime the bottom cells by telling them what they might be expecting. Your brain taps into memory and says, Listen, I've seen something like this before, and based on those experiences, and my understanding of the physical principles of the world, here's what I think is going to happen next. A Swedish researcher named David Ingvar was among the first to formalize this idea. He dubbed it "memory of the future."[17] It's what Gretzky had in spades. It's the key to the "one" CEOs that Horowitz values.

Information moving between upper and lower layers of the neocortex flows constantly in both directions—a never-ending loop of prediction, comparison, identification, and adjustment. In fact, the brain contains far more pathways for top-down clues than it does for bottom-up signals. The brain seems designed more for predicting things in the world than for gathering sensory data. We constantly make predictions, send them pinging out into the world, and learn something when they turn out to be right or wrong. Someone tells you you're going to meet Bill Clinton, and you predict what he's going to be like. You learn what he's really like by finding out which of your predictions were right and which were wrong.[18]

You experience this kind of prediction loop all day long, at a less than conscious level. As Grossberg put it, "When you walk downstairs in the morning and go to the refrigerator and open the door, you're primed by that series of events to expect certain things when you open the door."[19] If you reach for the orange juice, your brain automatically issues a series of predictions based on the many previous mornings when you've gone to the same refrigerator, opened the same door, and picked up the juice. Those predictions also factor in your experiences of the physical world that you keep stored in memory: what cardboard feels like, how it feels to pick objects up in your hand, how the muscles of your arm function, how gravity works. And your brain then measures those predictions against your actual experience.

To pick up the carton, you need to steer your hand toward it, successfully wrap your fingers around three sides, gripping with the right application of pressure—not too firm or you'll crush it. When you lift the carton, you have an idea in your head of the upward force required. This entire process happens at a level slightly below conscious thought, on autopilot. It works when the brain's predictions are highly accurate.

Emerging research is driving theories about why the brain works like this. In 2010 a Duke University team led by researcher Tobias Egner studied visual neurons by using brain-imaging technology. The team found "clear and direct evidence" that the brain predicts what it will see and tests the predictions against the images coming in through the eyes. What you "see" is something of a compromise between the two. "The visual cortex has assigned its best guess interpretation of what an object is, and a person actually sees the object," Egner reported. The Duke team said the research was significant enough to change the way neuroscientists study the brain.[20]

Predictiveness seems to allow people to selectively focus attention on really important tasks and filter out the ocean of sensory data constantly flowing in. "Consciousness is only a minuscule part of what we're doing," said Joaquin Fuster, a professor emeritus of psychiatry and neuroscience at the UCLA School of Medicine. When we talked to him, he was working on a book about how the brain makes predictions. "Ninety-nine percent of what we're doing in a given moment is totally unconscious. And it has to be that way. Otherwise you couldn't exist. It's only when something unexpected happens or when something ambiguous happens that we potentially become conscious of it."[21]

Another reason the brain works this way is that it lets people quickly identify and learn about unexpected things, which is a helpful skill in a complex and changing environment. Most of our everyday activities match our predictions, and that information simply isn't as important as the things that change. You already know how to handle the predictable stuff, whereas new situations might require a change in behavior—meaning they might require learning.

As Grossberg said, "Our brains enable us to successfully adapt, moment by moment, to environmental challenges."[22]

Could a company or other organization learn to act that way as well? A growing number, including Sisters of Mercy Health System, Sam's Club, and Caesar's Entertainment, are giving it a shot. As we will see later, predicting customer and competitive behavior will be a key differentiator in twenty-first–century business.

Paula Tallal is codirector of the Center for Molecular and Behavioral Neuroscience at Rutgers University and cofounder of Scientific Learning Corporation, which makes gamelike computer programs designed to enhance brain fitness in children. When she first got into brain science, she was working with adults who had lost their language abilities as a result of brain damage. "I was just absolutely amazed and horrified that you could lose the ability to communicate, to express yourself, or even understand what other people said," Tallal recalled.[23] That interest morphed into a career studying children who had difficulty talking or reading. Around 8 percent of children who have no discernable handicap struggle to learn language. Tallal engaged in brain research to try to find out why that happens and whether something could be done to improve the language capabilities of such kids.[24]

It's an interesting problem because language is so fast moving and so complex that processing language is the speediest your brain has to work. Your brain has to discern the tiniest changes in phonemes—for instance, distinguishing the sound of "add" from the sound of "at"—yet link those sounds to words, sentences, ideas, and the complex concepts they all represent. Your brain has to do all that in milliseconds or risk missing the next words in a conversation.

The problem Tallal identified is also interesting for our purposes, because if you look at her research carefully, Tallal has been studying talent backward. Instead of trying to uncover why an exceptional group is more talented than the norm, she has worked to find out why a lagging group is less talented than the rest of the population. In her studies, the talent is in language.

An important part of Tallal's research centers on that aspect of brain function called chunking. And chunking is important to predictiveness and talent.

Research has verified a physical property of the brain called Hebb's law, named after scientist Donald Hebb, considered the father of neuropsychology. The common shorthand version of the law is "Neurons that fire together wire together." But there's an added twist. Neurons that fire together nearly simultaneously in time wire that pattern together and then can replay the pattern equally simultaneously in time.

Experiences make neurons fire. So let's say a child hears the word "car" as a parent points to a car. The sound of the word "car" and the image of a car get encoded in neurons and connected to one another by axons. Repetition and prediction strengthen that bond. The next time the child sees a car, he's predicting—a fraction of a second ahead of time—that his parent will say "car." When the prediction is right, a ping of satisfaction rushes through the brain and the connection gets strengthened. Over time, the child experiences that many different kinds of vehicles are cars, that they have different properties, and that there are other words for cars, like "automobile." Each time knowledge is added, the pattern of connections gets more complex and refined. When a prediction about the concept of cars gets verified, that part of the pattern gets stronger. If a prediction doesn't match what actually happens, that part of the pattern gets adjusted or thrown out. This is, basically, how learning works.

The child builds a chunk—a complex pattern of information that fires at the same time—around the concept of a car. Hearing the word "car" or seeing a car or experiencing anything that has to do with a car fires up the chunk. His understanding of a car grows from the sound of one word and the sight of one image to a rich understanding of the concept of cars that comes to mind in milliseconds because all those neurons storing "car" data fire at the same time.

We chunk all sorts of stuff. We chunk walking. A toddler has to figure out how to stand and balance. Chunking that, balance becomes automatic and she can move on to taking steps. Chunk that, and steps become automatic and she can work on running. We chunk language. We chunk driving. (As long as nothing unusual is happening, most experienced drivers can zoom along at sixty-five miles per hour while barely paying attention to what they're doing.) We chunk our spouse's behavior, the way music sounds, touch-typing, shaving, and the way we do our jobs. We chunk our morning routines—like getting the orange juice out of the refrigerator. If not for chunks, we'd have to think about everything as if it were the first time. Chunks make us efficient. Bigger, more complex, more refined chunks make us even more efficient.

Predicting and chunking go together. A chunk feeds a prediction that's sent down through the layers of the neocortex and says, I've seen this pattern before, and this is what's supposed to happen. If the prediction is right, the chunk gets strengthened. If the prediction is wrong, the chunk learns. Your brain realizes that something about the pattern has to be adjusted, and it pays attention. The new experience gets added to the chunk, which tests the new pattern by generating a new prediction the next time. The predictions create the chunks; the chunks drive predictions; the outcome of those predictions improve the chunks . . . and round and round. If you have better chunks, you make better predictions. As you make better predictions, you build stronger chunks.

Predicting correctly is satisfying and gives you a safe feeling. The brain is predicting to try to reduce uncertainty, because uncertainty generally makes us anxious. Tallal suggests that this is a reason small children ask an adult to read the same book over and over. A toddler is still working on chunking individual sounds into words and words into grammatically correct sentences. Hearing the same sounds follow other sounds in a particular order, over and over again, teaches the child the sounds of language. By hearing the same words following other words in predictable sequences, combined with seeing the reinforcing pictures, the child gains the ability to predict what words come next, while also building up the rules of grammar for their language. Of course, this is happening when a child is listening to people talk, but the most predictable of all is when a story is read over and over again, or when a child learns nursery rhymes. When the child's predictions are true, it's satisfying, and the sounds start to wire themselves into chunks to create an understanding of language.[25]

This goes back to Tallal's question: why do some kids struggle with language even though they seem to have all the physical capabilities of kids who have no trouble learning language?

The answer lies in the speed of processing the sounds, and speed of processing turns out to have a lot to do with chunking. Research that Tallal and April Benasich conducted with infants in the first months of life showed that there were substantial differences in every infant's rate of auditory processing. When infants were presented with two different tones in different orders, some babies needed only tens of milliseconds (a millisecond is one-thousandth of a second) of silence between the two tones to respond correctly. Others needed as much as several hundred milliseconds.

This characteristic turns out to be important for building chunks for language because the acoustic difference between one speech sound and another can depend on acoustic differences that occur in less than forty milliseconds. If a child's brain can process information that quickly, he will be able to chunk the ongoing acoustic sounds of speech at a fine-grain level. If another child's brain needs hundreds of milliseconds to process the same information, that child will not be able to code the finest speech sounds into chunks. Instead, he will build only broader chunks of sound, such as syllables or short words, which occur over several hundreds of milliseconds.

Why would this difference in acoustic processing slow language development and create reading problems? The richness of a language comes from the ability to have a small set of sounds that can be combined in a very large number of ways to make an almost infinite number of unique words, sentences, and paragraphs. Instead of having to memorize every syllable or word in our language as a unique event, we just need to learn a few phonemes and then, through experience, learn the statistical likelihood of which phonemes most often follow others to form syllables and words, and which words are most likely to follow other words to form grammatically correct sentences that convey the meaning we intend. It's all about extremely accurate, fine-tuned predicting.

When you add that all together, Tallal's research on children's language abilities shows that a lack of talent in a certain realm corresponds to a relative lack of accurate chunking—which results in slower processing speed. Flip that around, and the kids who are relatively talented at language do more complex instantaneous chunking than kids who are not good at language.

So, why should some children end up with the advantage of faster processing speed than others? Research has suggested many potential factors that may influence each individual's processing speed. These include having a family history of language learning problems, being hearing impaired, or having a lot of middle-ear infections early in life, which can disrupt how fast sound gets from the ear to the brain. Processing the acoustic waveform of speech is the fastest thing the human brain has to do, so just listening to ongoing speech is important for practicing fast acoustic processing. The more speech a child hears, the more experience his brain gets in practicing fast processing. Many children simply don't have enough experiences with sound to build chunks and test predictions. Those kids aren't in homes where adults talk or read to them—often those families are in lower socioeconomic situations, where perhaps both parents are less educated, have to work long hours, and are rarely present. "The bigger the sample of language sounds, the more developed the patterns become," Tallal told us. "By the time kids from a lower socioeconomic family enter school, they can have a deficit of thirty million words that they haven't heard. The more you can predict, the faster you can process, and you need a lot of examples to feed those predictions."[26]

Chunking, then, is the creation inside the brain of superefficient mental models that process events in a flash—and it is a huge factor in talent among individuals. If those chunked mental models are responsible for talent in humans, some version of chunking or modeling ought to be a way to build talent into machines.

One further note about predicting and chunking: while predictions that prove right are satisfying, predictions that are *always* right become boring. Once you've chunked a pattern, it puts that activity on autopilot. As long as a prediction matches reality, you don't have to pay much attention. Yet every now and then, your brain wants stimulation or variety—and stimulation occurs when something lands outside your predictive capabilities. Novel, unpredictable events in our environment grab our attention and motivate us to learn.

A child may want to read the same book over and over, but once she's chunked that book and can predict everything in it, she'll want to move on to a different, perhaps richer book. A forty-year-old may find comfort and satisfaction in listening to the same music he's listened to for decades, but he buys new and different music that's outside his predictive model as a way to excite his brain. The desire for stimulation—these extrapredictive experiences—drives some people to vacation in foreign countries, try new foods, buy new products, seek new jobs, and even cheat on spouses.

Interestingly, the desired degree of extrapredictive stimulation varies from person to person. Some prefer the comfort of what they know—what they can predict—and choose safety more often than stimulation. They live in the same neighborhood for decades, go back to the same vacation spot, and buy the same brands time after time. On the other extreme are people we label as restless or adventurous. They crave having their predictions challenged. They're the ones downloading Tibetan monk chants because it's so different from the music they know. They're always traveling to far-off places, reading obscure authors, and are generally open to trying whatever is new.

Certainly, the level of desire to test predictability plays a role in developing talent. If someone is too safe in their choices of experiences, that person may not get enough new experiences to build a better, chunked model that would make more complex predictions accurately. But if one is too adventurous, one's experiences aren't repeated enough to be solidified into chunks that fire together and generate accurate complex predictions. Developing talent requires a balance of both—a desire to push past what's comfortable, yet a willingness to do the same things over and over to reinforce the links among neurons that make up chunks. Learning happens when predictions fail and our brains pay attention to something new. The learning becomes knowledge when predictions succeed and our brains reinforce and store the chunks.

Tallal's research shows that an ability to generate fast, accurate, complex predictions is a vital difference between children who are talented learners and those who are not. Predictiveness is the key to exceptional talent. People who are exceptionally talented in a field or ability make faster, more accurate and more complex predictions than the rest of us.

Over time, Wayne Gretzky built amazingly complex and huge chunks about what happens in a hockey game. He could see a set of movements in a game, and it would fire off a chunk that all at once would show Gretzky a picture of every dynamic in play. His brain would instantly recognize that he had seen this before and predict what would happen next. The prediction would often be right, and Gretzky would make a pass no one else would've seen as possible. The chunks allowed him to process the game so fast that he really did often understand what was going to happen before the other players. And because correct predictions allow the brain to process what's happening effortlessly, Gretzky's brain could put more of his play on autopilot than most players, allowing his attention to focus on subtle changes other players might miss, such as a little quirk in the way an opposing goalie moved.

For Ben Horowitz, a "one" CEO's brain operates in business the way Gretzky's brain operates on ice. The CEO chunks certain details about his company and industry, so he can see a situation and all at once know all the dynamics in play and make a prediction about what's going to happen. The chunks allow his brain to process everyday business in the background of his brain and pay attention to changes or opportunities others might miss. He can make such quick and certain decisions that it seems like he's reacting from instinct. But in reality, his brain is able to fire up a complex chunk of information all at once, so that he sees the answer in an instant—and is usually right.

As we've seen, one big contributing factor to the development of talent is training. Children need to hear millions of words over thousands of hours in order to build complex chunks necessary to process language quickly. Gretzky played hockey over and over from the time he could stand on skates. Concert pianists have played so much that they don't even think about notes during a concert, just about how the performance sounds and how the audience is reacting. In his book *Outliers: The Story of Success*, Malcolm Gladwell sets the standard of ten thousand hours as the amount of time necessary to become an expert. Practice a skill that much, and you chunk a mental model that can process that skill better and faster than most of the population. You become an expert or star in your field because you can see and predict in ways others can't. (If you're wondering, ten thousand hours is about three and a half years of eight-hour days, seven days a week.)

Oddly enough, despite the potential advantages of figuring out how to make more people more talented, only a few scientific studies have specifically looked at how talent works in the brain. Fewer still have looked into the process of prediction and its role in talent.

Here's one example of an empirical study that tried to uncover a correlation between predictiveness and talent. In 2005 at London's Institute of Cognitive Neuroscience, researchers looked at professional dancers from the Royal Ballet in London with different specialties: some were trained in ballet and others in a form of dance called *capoeira*, which is a fusion of dancing and martial arts that comes from Brazil. There was also a control group of nondancers.[27] The members of each group were scanned in an fMRI scanner while they watched videos of different kinds of dancing. The researchers then tracked the dancers' brain activity. The people in the control group didn't show a lot of brain activity when watching the videos, regardless of the style of dance. They might as well have been watching *The Simpsons*. Similarly, ballet dancers who watched *capoeira* videos didn't show a lot of brain activity—and vice versa. But when the dancers were shown videos of people dancing in the style they knew, their brains lit up like pinball machines. They were "dancing" the moves inside their heads, comparing the visual signals coming in to their own memory and predicting what was coming next.

The research so far leaves lots of unanswered questions about talent and the brain. How much of individual talent is the result of "hardware"—a genetically well-wired brain? How much is "software"—the experiences and learning that create chunks? Horowitz thinks that, in business, ones and twos can each learn just as much about a company, but ones still operate at a higher level. "It's hard to take a two and make a one," Horowitz said.

Yet as Tallal found, to some degree talent can be developed, not just born. So if people can learn to become talented, perhaps machines can too.

There is a good reason why there hasn't been a lot of empirical scientific study around the predictive nature of talent. It would be difficult to, say, put superstar quarterback Peyton Manning in an fMRI machine and observe his brain activity while he's making predictive decisions during a football game. Or get a superstar CEO to lie in one while he's making decisions in his office. (However, this might change. In late 2010, the University of

California's Swartz Center for Computational Neuroscience announced it had invented a "mobile brain/body imaging modality" that could capture brain activity "as subjects actively perform natural movements."[28])

Still, some interesting theoretical work helps shed light on predictiveness and its role in talent.

At Columbia Business School, William Duggan came up with the concept of *strategic intuition* while studying Napoleon.[29] Duggan went back to the classic book *On War* by Carl von Clausewitz, first published in 1832. Von Clausewitz described how Napoleon won his first battles at the age of twenty-four—when he was a bratty little general whom none of the other officers took seriously. Napoleon had almost no battlefield experience, but he'd studied military history extensively, especially details of how battles were won and lost. Napoleon loaded all of that information into his head, along with everything he could learn about his own army, his opponents, and the geography of the battlefield, and then he marched his men off to fight *without much of a plan.* Napoleon had no more of a predetermined plan for how he'd fight a given battle than Gretzky had for how he'd score a goal. Instead, Napoleon watched events unfold and had the presence of mind to see them from a high strategic level, and he'd wait for a flash of insight, or *coup d'oeil,* as von Clausewitz described it.

On a battlefield in the 1790s, when people and information traveled slowly, perhaps this wasn't a two-second advantage but more of a two-hour advantage, or even a two-day advantage. Nonetheless, Napoleon was able to take in events and make on-the-spot predictions about what would happen if he took a certain action. He could do this better and faster than other generals and won battles that way. In Napoleon's first Italian campaign, his army of 35,000 faced two enemy armies, each with 35,000 troops. In other words, he was outmanned two to one. If Napoleon had stuck to the French army's original orders, he would've had to fight a force double the one he commanded and surely would've lost. Instead, Napoleon saw and exploited an opening: he marched his army between the other two forces, turned and defeated one, then turned and defeated the other.

In reading von Clausewitz's account, Duggan said, "It occurred to me that it seemed an awful lot like modern research on expert intuition." He connected it with the latest thinking in neuroscience, which led him to write a book, *Napoleon's Glance.* Duggan defined the idea of strategic intuition this way: it was "the selective projection of past elements into the future in a new combination as a course of action that might or might not fit previous goals."[30] In other words, it means predicting what will happen based on immediate events and taking action based on that prediction. Sounds a lot like Gretzky on ice or a "one" CEO in a tech start-up. Everyone makes these predictions to some degree, but the most talented among us simply do it better and faster than everyone else. Napoleon was the Wayne Gretzky of eighteenth-century European warfare.

Gary Klein of Applied Research Associates is one of the best-known thinkers about how people make decisions—especially split-second decisions in times of stress or action.[31] Lately, he's zeroed in on what he calls *anticipatory thinking.* Again, it's something we all do, and we get better at it as we learn more and chunk things together so we can put them on autopilot. New drivers get in more accidents than experienced drivers because they're paying too much attention to the basics of driving and don't yet have a mental model for the patterns that signal trouble on the road, according to Klein. As drivers gain experience, much of the process of driving is done on autopilot and "they have heightened sensitivity to weak signals that would be ignored by those with less experience," Klein wrote. Experienced drivers see events that are unfolding, anticipate what will happen, and take action to stay safe. "They have an edge in detecting problems," Klein wrote. And that edge is anticipatory thinking. Again, the better someone is at anticipatory thinking, the better he or she performs, Klein concluded.[32]

Klein adds two important thoughts to the conversation about predictiveness.

First, people who are talented don't just make predictions based on events; they also make predictions based on a *lack* of events.

This means they catch the notes that didn't get played because someone in the orchestra missed them or recognize the deal that didn't happen or the move an opponent didn't make. It's much more subtle than processing the things that do happen and takes a greater level of knowledge and a higher level of thinking. But this will prove particularly important when we discuss instilling predictive talent into computer systems. Databases can't process nonevents. If something didn't happen, the data doesn't get created, and if there's no data, today's software can't make use of it, even though there is valuable intelligence in making sense of things that don't happen. A simple example: if someone goes on an e-commerce Web site, looks around, but never registers or buys anything, that information disappears. The site doesn't recognize that someone who could've become a customer came in and left, or why. If computers are going to become predictively talented, they'll have to be able to read and understand nonevents. That would be a profound change from the way information technology works now. At a corporate level a company like Xcel Energy could look for absence of calls to know that an outage has cleared up on its own.

Klein's other important thought is that anticipatory thinking doesn't just happen in individual brains. "We believe that anticipatory thinking is critical to effective performance for individuals *and for teams*," Klein wrote with two coauthors, David Snowden and Chew Lock Pin, in a book he was working on when we talked to him (italics added). If anticipatory thinking—or predictiveness, or strategic intuition—can occur in teams, it should be able to take place in larger organizations. The trick is creating systems that make that possible.[33]

Two concepts related to predictive talent have made their way into popular culture since the mid-2000s. They are *blink* and *flow*. Malcolm Gladwell's 2005 book, *Blink: The Power of Thinking Without Thinking*, explored human intuition. Gut instincts are often correct, and sometimes better than thought-out, analytical decisions, Gladwell found. And people can get better at intuition. The people Gladwell described in *Blink* "are very good at what they do and . . . owe their success, at least in part, to the steps they have taken to shape and manage and educate their unconscious reactions. The power of knowing, in that first two seconds, is not a gift given magically to a fortunate few. It is an ability that we can all cultivate for ourselves."[34]

The concept of "flow" was proposed and popularized by research psychologist Mihaly Csikszentmihalyi. It describes the state of complete immersion and total focus, when time seems to slow down and everything goes right. It's the absolute optimum human mental state. When in sports someone is having a great game, he or she is probably in a state of flow, losing all sense of self-consciousness and time, simply letting instinct take over. But it's not limited to sports—a businessperson giving a great presentation could be in a state of flow, or a surgeon completely immersed in the operation at hand.

Flow seems to be a state of pure predictiveness. The person in a state of flow is firing complex chunks so rapidly and seamlessly that she feels like she can see what's going to happen before it happens and can act with complete certainty and confidence. It's easy to be confident when you know how your action will turn out.

Later in the book, we'll look at how brain research and theories about predictions and talent apply to talented individuals. And we'll explore how the same concepts can work in machines and enterprises. But first let's look at how some of these traits come together in the cooking of a famous chef known for making crazy concoctions.

Elizabeth Falkner is one of the top chefs in San Francisco. She started Citizen Cake as a small bakery and grew it into a successful restaurant. She followed that with Orson, a critically acclaimed restaurant in the trendy South of Market neighborhood that serves bistro-style food in a hipster setting. (The restaurant names are nods to Orson Welles and his film *Citizen Kane*—and a reflection of Falkner's film school background.) Falkner is a fixture on the Food Network, a spiky platinum-haired woman with a punk aesthetic, usually wearing ripped jeans and a rocker T-shirt.

She is known for putting together ingredients that don't seem like they could—or should—go together. There's her orange-infused fettuccine, her fried chicken and kimchi, and her beef short ribs with ricotta gnudi. She created a dessert that is essentially a Monte Cristo served with smoked almond ice cream. Her dishes win her awards and a steady clientele.

We asked Falkner to help us try to understand how she can come up with dishes like those and *know* they'll be good.

To explain her thought process, Falkner dissected one of her recent creations, an East-meets-West dessert of red adzuki bean ice cream served with soy caramel and yuzu fudge sauce, genmai Rice Krispies treats, and an agar and red bean gelée.

Falkner created the sundae when she had a deadline to meet for a Japanese "World of Flavor" demonstration at the Culinary Institute of America. Her mission was to use certain Japanese ingredients but appeal to a sophisticated American palate. As usual, she wanted to create something that would defy expectations. "What's happening in my brain is, I'm interested in articulating the flavors with a sketch or a painting that people can wrap their heads around," she told us. "They know an ice cream sundae, and this will look like one. Yet it'll have all these unfamiliar textures, all the salt and sugar, and people will say, 'Wow, I've never had anything like that before.' "[35]

When she gets an assignment like that, her first goal is to carve out some time to *think*. "I like to really sit and imagine the textures and flavors that I want," she said. In this case she drew on valuable experience she gained at her first big restaurant job as the pastry chef at Elka in San Francisco's Japantown in the early 1990s. There she learned a great deal about Japanese ingredients and techniques. Drawing on that experience, she started creating her unconventional sundae with the idea of the red bean paste, which she has had in various Japanese desserts. "I know it's going to be like squash or chestnuts or something like that," she said. "I know it'll give the ice cream a really nice texture because of its starch."

She continued: "I know I'm going to put in milk or cream with the beans, and I'll have to see how sweet they are." She considered experimenting with fresh and canned beans, as she wasn't sure what the supplier would ultimately provide at the Culinary Institute Japanese food demonstration. Her goal, she explained, was an ice cream that was not "super sweet and fluffy."

"I know soy sauce goes really good with caramel, because of the salt in the caramel," she said. She has long been intrigued by the popularity of certain candy bars, like Snickers, which use salty peanuts to highlight the sweetness of chocolate and caramel. Similarly, she said, yuzu, which she describes as a cross between a lemon and a tangerine, "will be good with chocolate, because it has such a fantastic citrus flavor. And there are other little things I'll do that I know will help make other notes within the dish."

Genmai, for instance, is a green tea with puffed rice in it, which gave her the inspiration to Japan-ify the classically American Rice Krispies treat.

By that point in the process, Falkner had become certain of her creation and how it would taste, even though all of this happened in her head, in a flash of inspiration. She was drawing on her extensive knowledge, but not in an encyclopedic way—she didn't have time, with the deadline approaching. She thought the assignment through using a mental model—a series of chunks—she'd created in her head about desserts and flavors. She didn't have to think about every individual property of every ingredient—she could leave all that data in the background and take it to a higher level.

"I only brainstormed about it this morning," she told us. "I have a sense of urgency to get my list of ingredients to the school. This is exactly how I do all my dishes. I'll have multiple things going on—and I do dream about them too—but I know exactly what that has to taste like. I have a good taste memory. I'm still crafting it in my head. How much cream should go in that? No, not too creamy. How much yuzu in the chocolate? I don't want to overpower the ice cream. That particular component may not work. I want to figure how the chocolate can work in this dish and not overpower the whole thing."

But how did this model blossom in her head in the first place? Falkner's decision-making process is a version of the Wayne Gretzky story. Her brain seems to be naturally wired in a way that's perfect for artistic cooking, and she got a ton of experience—those ten thousand hours of practice—that let her chunk together a lot of knowledge. Either one by itself would not have created a Falkner. But forged together through the circumstances of her career, they allowed her to become the famous chef she is today.

Gretzky was helped by what appeared to be a handicap—his size. Falkner was helped by the fact that she wasn't trained initially as a cook.

Growing up, Falkner focused on the arts. Her father, the abstract painter Avery Falkner, is an art professor at Pepperdine University, and his works dominate the walls of Elizabeth's restaurant. One brother is a rock musician, and one is a storyboard illustrator, actor, and dancer. She painted a lot in high school, and her dad gave her an expensive camera that fostered an interest in photography. She went to the San Francisco Art Institute to finish college and study experimental film. She worked in restaurants initially only to support herself. "I got into the food thing because it was big in San Francisco in the late eighties and early nineties," she said. "I always cooked, but I had never thought about it as a career."

In high school, Falkner worked at Bud's Ice Cream, where she developed new flavor combinations, and at a Bay Area Italian deli that baked its own bread for sandwiches. In the 1980s, inspired by Famous Amos, Mrs. Fields, and others, Falkner began making her own cookies. She took a job at the original Williams-Sonoma store at Sutter and Taylor streets in San Francisco, where shoppers included chefs Julia Child and Marion Cunningham. San Francisco was exploding with great chefs and restaurants. She took a job at tiny Café Claude, a French café on an alley near ritzy Union Square, and revamped the menu, which had no house-made desserts. Soon she was baking bread pudding, and—because no self-respecting French café should be without one—an apple tart. In short order, she was made the head chef and jettisoned her film career.

One morning, an expediter for Masa's, one of the top restaurants in San Francisco, while eating at Claude, told Falkner of an opening at Masa's. Falkner got the job, working under famed head chef Julian Serrano, some nights peeling two cases of pearl onions to get them to a uniform size and slicing her fingers to ribbons in the process. She soaked up food culture, learning as much as she could, eating everywhere—including, fatefully, Elka, where Traci Des Jardins and Elka Gilmore created a blend of French and Asian cuisines. Falkner told them that their desserts weren't working. So they hired her.

"She wasn't proven. She'd been an assistant. But there was some glimmer of basic innate talent we saw in her that made it clear she was going to go someplace," Des Jardins recalled. Falkner's first dessert for Elka was inspired by her movie background. She named it "The Battleship Potemkin" after a 1925 silent film by Sergei Eisenstein. The dessert featured small Odessa Steps made out of chocolate shortbread and raspberry sauce representing the bloodshed of the civilians slaughtered on those steps in the film's depiction of the Russian revolution. "It was all black and white and red," she said. "I told the servers about it, and they all looked at me and said, 'What?!' "

When Des Jardins opened the restaurant Rubicon in San Francisco's Financial District, Falkner went with her. After three years at Rubicon, Falkner's side business of baking birthday and wedding cakes was taking off, and she opened Citizen Cake. It evolved from bakery to lunch spot to full-service restaurant and bar. She opened Orson in 2007. By then she was a star chef, a whirlwind who was constantly concocting all manner of surprising dishes in her head.

But does she have that intuitive predictiveness of a Gretzky? Or the situational awareness that leads to the brilliance of a Napoleon?

Gabriel Maltos, one of the waiters at Orson, often fields requests from people with one allergy or another. "We'll have a lot of people come in with special needs," he told us, "and she'll tinker with it, and it'll go out, dairy-free and beautifully done." The key, Falkner said, is "if somebody has an allergy and you have to twist it around, it's being so familiar with the medium. Certainly I do make decisions with artistic license, but a lot of that license is based on the foundations that I have."

Added Maltos: "She's got that natural sense. Some people go to culinary school for years and never get it." Others can have the sense, yet never refine it.[36]

Which brings us back to the question, How much of exceptional ability is hardware (the raw genetic wiring and processing

The Talented Brain

Eduard Schmieder was sixty-two years old when we interviewed him in the bright sunroom in the back of his home in Los Angeles.[37] He had become perhaps the best violin teacher in the world and a conductor who had performed at Carnegie Hall in New York and Disney Concert Hall in Los Angeles. Yet those achievements obscured the fact that Schmieder today might be known as one of the world's great violinists had Soviet operatives not, apparently, beaten him nearly to death in the 1970s. We say "apparently" because Schmieder won't talk about the episode that changed his life, referring to it only as "my injury." But we can piece together the circumstances from other accounts. In the heat of the cold war, Schmieder had become a high-profile *refusenik* in the USSR. The *refuseniks* were generally Jews who applied to leave the Soviet Union and were not only denied but then considered traitors for trying to leave. Schmieder was no doubt a particular problem—a star musician who, by becoming a *refusenik*, turned into a visible, public symbol of communist oppression. One way to stop him from being so visible was to stop him from playing. The Soviet-imposed "injury" that Schmieder suffered left him paralyzed for nearly two years and unable to ever play the violin as he had before.

This made Schmieder more aware of how his talent worked—because after the injury, he could no longer use his talent the same way.

Schmieder was born in Lviv, in the Ukraine, which at the time was part of the Soviet Union. He was an only child, and when he was very young a friend of his father's visited the house and played a violin. Schmieder begged his father—who was not particularly musical—to buy him one of the instruments. Somehow, to the father's amazement, the boy and the violin clicked. By the time Schmieder was eight years old, he was performing onstage—he was a child prodigy. He didn't have to work at it much. He loved to play, and it just came naturally to him. Later in life, he says, he and a legendary classical violinist got talking about how neither of them ever practiced much or ran through scales. "I could never understand why one needs to do exercises until I had my injury and had to start from a position of disadvantage," Schmieder told us in his Russian-accented English.

When he performed, he felt like another power took over him. "When I was a child I remember coming onstage and from this moment, I was not myself," he said. "Something was happening to me, and it came from somewhere else and I did not think about anything. The minute I start performing onstage, I can't think anymore. I am not myself anymore. Somebody is doing this for me."

Schmieder could instantly and consistently will himself into the state of flow described by Csikszentmihalyi. That's what happened every time he stepped onstage to play. He had something much more raw than anticipatory thinking or strategic intuition—both of which rely on high degrees of knowledge and experience that then get chunked to form a tidy, efficient mental model. By age eight or ten or twelve, Schmieder couldn't have soaked in enough information to build that kind of model. He didn't have Gladwell's ten thousand hours of experience—nearly four years of eight-hour days—at such an early age. Yet he had astounding predictive capabilities in music—the very essence of talent. "The music somehow goes in front of me," he said. "It's more than just experience; this is a born ability. I hear it before I produce it." Or at least he used to, before the injury.

Schmieder at a very young age could predict what set of actions would create a certain kind of sound—including the emotional content of the music. He could make those predictions better and faster than most people who train on the violin for decades. That's why he could beat out experienced violinists for gigs and positions in orchestras.

Schmieder's explanation for his untrained talent is a bit mystical. "I believe our genes not only pass on physical traits but have emotional memory. These memories keep accumulating through tens or hundreds of generations and somehow manifest themselves by passing on knowledge that cannot otherwise be gained in a physical lifetime," Schmieder told us. "My subconscious mind is somehow giving out this emotion that's accumulated for such a long time, but my conscious mind does not interfere. I have to rely completely on my subconscious mind, which has a connection with some other energy, which somehow is helping this (music) capability to work. I don't know how."

Mystical, yes. But in a way, Schmieder is on to something.

And his experience has a connection, believe it or not, to autism. This connection gives us clues about how to build innate predictive talent into machines and organizations.

In 1987, thanks to a BBC special, the world learned about Stephen Wiltshire, then just twelve years old. He is one of the purest examples of savant syndrome of the past few decades—a far more extreme version of the young Eduard Schmieder being able to play the violin without much training. Wiltshire was diagnosed with severe autism when he was three, and he did not speak until the age of five. Even today, as an adult, he can barely interact with people and cannot perform most of the tasks that make up ordinary life. Yet Wiltshire became famous because he can do something literally no one else can do. He can see an entire cityscape once and then draw it perfectly.[38]

This started when his London grade-school class visited ornate Albert Hall. After seeing the building, Wiltshire returned and drew a detailed, exact reproduction. He soon began drawing buildings and skylines, and after his initial BBC fame his drawings were featured in books and calendars. In 2001 the BBC flew Wiltshire over London in a helicopter. Back on the ground, Wiltshire drew a bird's-eye view of the city, with more than two hundred buildings to perfect scale, in three hours. In the United States, he was taken on a twenty-minute helicopter ride over Manhattan and then drew a perfect twenty-foot panoramic view of New York from memory. He even gets the number of windows correct on buildings and includes architectural details barely noticeable to most people.

Amazingly, Wiltshire doesn't sketch first or plan out size and perspective—he just starts drawing, working from one end of the city to the other. It's as if his mind takes a video of what he sees and stores it in memory, which he can then watch as clearly as if you or I shot a video with a camera and played it back on a screen. And in fact Wiltshire once told the New York *Daily News*, "Everything is like a TV show—I have never drawn from a sketchbook." Wiltshire simply directs his hand to make a copy of the video image in his head—kind of like if you memorized the Gettysburg Address and then directed your hand to write out the words. To Wiltshire, drawing the cityscapes is that simple—while something like going to the grocery store would be unimaginably difficult.

Over the past couple of decades, autism has been studied intensely—and there's been a particular interest in autistic savants. Researchers want to figure out how they do what they do. And there's often a subtext—the question, How can *the rest of us* do what they do? There's an implicit idea that being able to draw from a photographic memory, calculate large numbers in a second, or reproduce music heard just once is a higher form of brain power. The belief is that we would be better and more productive people if we could unlock or create savantlike capabilities within ourselves.

But actually, that idea is badly misguided.

At the Centre for the Mind in Sydney, Australia, researchers Allan Snyder and D. John Mitchell have been studying the brain processes of savants for more than a decade. They explain that people with autism focus on the trees rather than the forest. Savants with autism waste no brain power at all on seeing the forest—to them, it's just not there. Instead, all their brain power zeros in on the trees, in minute detail, giving them "privileged access to lower levels of raw information" not typically available to the rest of us.[39]

Let's go back to our description of how the brain works. The lowest layer of the neocortex collects raw information in granular detail from our five senses and passes it up to the next level, which starts to put the pieces together to form an image or sensation. As the information moves up to yet higher layers of the cortex, it's assembled into broader, more sophisticated, more creative concepts. At the highest working levels—Gretzky on ice—the details are left behind, chunked into an efficient mental model that can process what's happening in a flash of recognition and anticipate the near future with incredible speed and accuracy. Those predictions—not the details—create our reality.

As powerful as our brains are, they don't seem to be powerful enough to record and recall every tiny detail *and* process high-level concepts. Given a choice, there's greater benefit in the high-level concepts, so our brains focus on those. We get very good at using the details we sense to create the high-level concepts, and we're very good at knowing what details to essentially forget—or store somewhere in the brain's attic for use later if we really need them. As Snyder said: "What matters for survival is that we have a concept we can work on—it's a face and it's friendly, say—not a mass of detail about how we arrived at that conclusion. So in normal people, the brain takes in every tiny detail, processes it, then edits out most of the information leaving a single useful idea which becomes conscious."[40]

Other researchers go a step further and conclude that normal brains are so focused on high-level cognition, we don't ever store the vast majority of details we take in—our brains use the details to get to a high-level thought and then fling them overboard. At our best, we are extremely efficient forgetters—to our benefit.

Imagine, though, what happens in a highly intelligent brain of an autistic person. For reasons not entirely clear, an autistic person's brain can't get to the higher-level processes. It can't put details together to understand the broader concept of human interaction or living in a city. In the case of an autistic savant, the brain grasps only the details—the trees, not the forest—and focuses nearly all its processing power on collecting and recalling them. There is no reason to forget things or relegate them to the basement because an autistic savant brain doesn't need capacity to do higher-level thinking—it can devote everything to details. Autistic savants' existence is made up of massive amounts of data and little understanding. They *can't* forget. Which is why Stephen Wiltshire can draw a city from memory. To the rest of us, Wiltshire's talent seems amazing and worthy of our envy, yet the very act of not forgetting burdens Wiltshire's brain to such an extent that he can't think a higher-level thought.

Savantlike capabilities would be great to have—if your brain could process both the details and the high-level thinking. But it can't. Scientists have been doing experiments with transcranial magnetic stimulation—electromagnetic pulses fired through normal human brains. TMS tends to interfere with high-level thought and opens pathways to details. A *New York Times* writer, for instance, tried it and was temporarily able to draw cats with much more accuracy.[41] Yet that seems to be turning evolution backward. Gretzky's talent is not remembering every detail of every game he's played—it's using those details to instantly create anew.

Consider this: newborn babies are basically savants. That's a reason they can learn language so well and so quickly. They don't do much high-level thinking at that stage of life, so their brains can focus on absorbing and remembering combinations of sounds in astounding detail. As they grow up, they move on to higher-level thoughts—and learning a new language becomes a difficult chore. Since our brains have to make a choice, the higher-level thinking is far more valuable than collecting and recalling tons of detail. Savantism is moving backward. Forgetting is evolving.

As we'll explore later, this is a fabulous idea for technology. All of computing history so far has been aimed at gathering more data, storing it, recalling it, and analyzing it. Computers are, really, autistic. And that's a big reason they've been so useful to humans. Computers have been made to excel at exactly those things humans are bad at. They can store and recall every detail or do calculations in an instant. Most humans can't do those things because we spend so much energy thinking about higher-level concepts. So it's been a beautiful, symbiotic relationship.

Yet if we want computers to be a little more human—to be a little bit talented—this business of collecting and delving into mountains of data eats up too much processing and storage resources and ultimately takes too much time. To take machines to the next level, we're going to have to teach them how to forget.

There are two ways to be funny. One relies on a mental database; the other, on an efficient mental model.

Joan Rivers is a classic database comic. At one point in the documentary *Joan Rivers: A Piece of Work*, released in 2010, she's in her home, standing in front of a wall of filing cabinets storing thousands of three-by-five index cards. Rivers explains that she thinks of jokes constantly and writes them on whatever is handy, eventually typing them on index cards and filing them by topic. One individual drawer, for instance, is labeled, "NEW YORK: NO SELF WORTH." The wall of files is, essentially, Rivers's master database.

When she's about to do a stand-up gig or appear on a talk show, Rivers goes searching in the files for appropriate jokes. Some of them are timeless: "The one thing women don't want to find in their stockings on Christmas morning is their husbands." Others might be topical, perhaps making fun of a celebrity popular at the time: "Elizabeth Taylor has more chins than a Chinese phone book." For those, she might cross out the old celeb's name and write in someone more up to date. Once Rivers has assembled enough jokes for the occasion, she memorizes them—pulls them from the master database and puts them into her brain's equivalent of computer memory, ready for instant access. Then, in her performance, she relies on recalling and delivering those jokes perfectly.

Joan Rivers, of course, is a very successful, enduring comedienne. As a young woman in the 1960s, she did stand-up at Greenwich Village comedy clubs before landing appearances on *The Tonight Show* and *The Ed Sullivan Show*. By the 1970s, her career had taken off, and over the next thirty years she became a talk show host, movie actress, Las Vegas stand-up regular, and Oscar night red-carpet show host. Through it all, she owed much of her success to that huge database and an inherently slow process. It's telling that in *Joan Rivers: A Piece of Work* Rivers says that she doesn't consider herself a classic comic. She thinks of herself as an actress playing the role of a comedienne. Take away her lines and she'd be lost.

If a group of software programmers today were to set out to build a computer comedian, they'd probably model it after Rivers. They'd create a giant database of jokes, plus rules about when to pull out what kinds of jokes and how to sequence jokes into a routine. Such a computer might even pull off a pretty good stand-up act. But would the computer *actually* be funny? Could it pull off improvisation? Could it unleash a witty zinger in response to a talk show host's question? Not likely.

Now, compare Joan Rivers to Mo Rocca, who has made a career of the instant, in-the-moment, improvised quip. He has never been as big or successful as Rivers, but you might've heard Rocca on his frequent appearances on NPR's *Wait, Wait, Don't Tell Me* or seen his work on CBS or VH1. He comes off as something of an intellectual Pee-wee Herman—a little goofy, a little nerdy, yet smart as hell. He has a killer comic instinct. Rocca can take in what's happening, see an opening, craft something on the spot, and drop it into the conversation with perfect timing. It's a completely different model from Rivers.

So we asked Rocca how he does it. And he began to dissect a particular moment during one of his appearances on CNN's *The Joy Behar Show.*[42]

First, some background. Rocca, born in Washington, D.C., in 1969, as a boy attended the exclusive Georgetown Preparatory School. Like a lot of eventual comedians, he got attention by seeking laughs. "One of the big challenges was making the teacher laugh—then you'd really done it," Rocca told us. "If your one-liner, if your little unplanned moment, perfectly sprang on a teacher and landed just right, then that was pretty cool." By sixth grade, he said, "I realized I was very good at saying just the right thing at the right time." He offered no great insight into *why* he was particularly good at this. As the cliché goes, it came naturally to him.

Rocca went to Harvard, got involved in theater, and became president of the school's Hasty Pudding Theatricals. After graduating with a BA in literature, he moved to New York and tried stand-up—and found he wasn't very good at it. "Stand-up comedy is a one-way conversation," Rocca said. "I admire the people who do it well, but I'm better in a dialogue."[43] Stand-up relies on a database. Rocca had a different system in his head—an anticipatory mental model. He increasingly followed that muse, which led him to more improvisational gigs on *The Daily Show with Jon Stewart* and *Wait, Wait, Don't Tell Me.*

It's not that Rocca doesn't prepare. He researches topics and writes jokes and practices them in front of a mirror. But he's at his best when he pushes that to the background and lets his predictiveness take over. Instead of focusing on the jokes themselves, since grade school he's focused on refining the rules that tell him when to insert a comment and how to instantly craft the right thing to say. He's not sure what all those rules are, but he can articulate some. For instance, listen first and find a connection with the other people in the conversation. Or, do the unexpected: in one segment about NASCAR, Rocca played against type and came off as highly knowledgeable about car racing.

"What I do is, I'm looking for that opening to land the perfect little one-liner," Rocca said. "I listen, and I wait, and we're all building the moment as we speak, and I see an opening right here and I think, I'm going to put the cherry on this sundae right . . . now!" He paused a second and added, "Of course, sometimes the cherry rolls right off."

By the mid-2000s, Rocca had become well known for his wit. And that landed him on the set of Behar's show in May 2010, along with stand-up comic Colin Quinn. Behar and Quinn had known each other for years. Rocca barely knew either of them. The three of them got into exchanges typical for the show, riffing on news and celebrities. An item came up about Sarah Ferguson, formerly married to England's Prince Andrew.

Quinn barked, "Who?"

Rocca: "And you know she did Weight Watchers. She did a lot of stuff, so in a way I respect her a lot."

Behar: "Yes, but she spends a lot, too."

Rocca: "Yes, and for that she should be beheaded."

(Usually, these things are funnier on video than in print.)

As the show wound down, Behar turned to her old friend Quinn and said, "It took me a long time to get you on this damn show, Colin. I've known you for twenty-five years."

Quinn responded that it was payback for Behar never coming on his show on Comedy Central. Quinn and Behar started bickering about it.

"And I'm sitting in the middle of them, and they start going at it," Rocca recalled. "Instantly, I thought, Okay, I know I want to land something right here. I need to wait long enough that the joke really lands, but I don't want to wait too long that the moment passes.

"And so while they're yelling over each other, I threw my hands up to my ears and I screamed, 'Mom! Dad! Stop it!' "

Rocca continued: "And it worked. But what's so funny is it all was happening in slow motion, because I remember thinking, Okay, they've started to bicker, this is perfect. If they can just bicker for about a second more, then it will be long enough that the joke of me as their child, as a child of this man and woman shrieking, will be really funny. But if I wait too long, it'll pass or one of the other two might land a joke to end it.

"And that's really an example of what I've tried to do all my life," Rocca said. "It's as much instinct as calculation. When you're in the zone and really grooving, it's like this."

This isn't a judgment about whether Rivers or Rocca is funnier. It's just that they use different models. Rivers relies on massive amounts of information assembled in the past, and that model works wonderfully when there's ample time to search and deploy that information. Rocca assembles information too, but his brain chunks it and uses it to create an agile mental model that can make predictions in real time, based on what's happening amid the flow of events.

If you put it all together, Rocca is not that different from Gretzky. Rocca created an efficient, fast-acting, predictive mental model in his brain. On the set, he gets into a state of flow, uses his senses to take in the developing situation around him, and is able to see it in slow motion—or to put it another way, see what's happening faster than everyone else. He is two seconds ahead of the conversation and can predict what to say and when—and be right most of the time.

Like Ben Horowitz's "one" CEOs or a young Eduard Schmieder or savant Stephen Wiltshire, Mo Rocca's talent came to him naturally and early. He didn't initially get talented by rigorously developing his improvisational humor—rather, he first was talented at it, and then he developed that talent further. We can learn a lot from trying to understand how such people do what they do, but it can be a little frustrating as well, because we can't figure out how they came to be *able* to do what they do. Schmieder offered his theory about memory getting passed along in DNA. Most observers, even neuroscientists, just say that such naturally talented people have brains that are "wired" differently from the rest of us. But nobody's really sure what such "wiring" means.

"Some people's brains are by nature better suited to certain tasks," Jeff Hawkins told us. "In normal humans, the size of region V1 in the neocortex varies by a factor of two or three. People with larger V1s—the first area of vision—seem normal, but they have higher visual acuity. Similarly, we can imagine that some people's brains are by nature better suited for math, music, language, or hockey."[44] Hawkins added that no doubt Wayne Gretzky's brain "was better suited for the kind of tasks hockey requires." Yet scientists are unsure why. Was the physical structure inside Gretzky's head, from the time he was born, all that different from the structure inside the head of an average recreational-league hockey player, or even an NHL player who is good but not Gretzky-level good? The question hasn't been answered yet, and in fact there have been few studies that have even tried.

We do know about one factor that seems to play a role in highly talented people: their brains become supersynchronized when they're using their talent. George Mason University neuroscientist Jim Olds explained that the brain can work like a symphony orchestra. When not focused on a task—say, when you're in the shower and letting your mind wander—your brain is like an orchestra warming up. All the instruments—or in your brain's case, neurons—are playing something, but they are not in sync with any other instrument, so it just sounds like noise. Now imagine that the first violin starts playing a melody that another musician recognizes, and that other musician joins in. Soon a few more musicians join in, and then more, and spontaneously out of the noise a song emerges. This goes on until a musician across the room realizes the song reminds her of a different song and starts playing that one. Other musicians hear it and decide they like the new song better, and they start playing it—until the same thing happens again and everybody switches to yet another song.

You might recognize, in such a free-flowing, free-association ebb and flow of music, something similar to the way your thoughts ping around in the shower or while driving down a familiar highway. Your neurons are all firing with their individual ideas, and as long as that continues, you've got a lot of random noise going on in your head. But when a group of neurons decides it likes the idea coming from one place, other neurons join in, contribute, and recruit yet other neurons to join in too. The new neuron recruits in turn recruit others, and suddenly your brain is synchronized, i.e., thinking deeply about something—until a different neuron has an idea that recruits a wave of neurons its way, and your brain shifts to that thought. Over and over again, with magnificent speed, your brain bounces from topic to topic. If you're trying to focus on work, those bouncing thought patterns can be distractions.

When is an orchestra at its best? When a skilled conductor comes in and gets the whole orchestra to play together for an extended period of time. The instruments don't play exactly the same notes. That would be *too* much synchronization. (Olds said that's what happens in a brain during an epileptic fit: oversynchronization.) Instead, the notes played are all coordinated for a greater effect. Similarly, a human brain is at its best when the neurons coordinate for a specific task for an extended period. In fact, that's exactly what happens when someone gets into that state of flow we discussed earlier. The whole brain is completely on task. There are no neurons in the corner trying to recruit other neurons to think about what restaurant to go to later. The neurons are working together to play the same song.

How do we know this much? Because of experiments done on Buddhist monks. The monks can will themselves into a state of flow—which they would call meditation. The handy thing about a Buddhist monk meditating is you can put him in an fMRI machine and watch his brain work. "Their brains settle into complex rhythms, but very coordinated," Olds told us. "It shows aspects of a symphony orchestra. It allows many brain regions to communicate easily with each other—they interact as if they have a clear channel to each other."[45]

This is at least a partial clue as to how talented brains are wired. Talented people seem to be born with an unusual ability to focus the brain's resources on one task. Savant artist Wiltshire is an extreme case. People like Gretzky or Rocca can will themselves into that state of coordination, or flow. They seem to start up their mental models, quiet everything else, and open channels between regions of the brain. They run their minds so efficiently that time seems to slow down, possibly because they're actually perceiving the world faster than the rest of us—which helps them get their predictions out ahead of the rest of us.

Aside from the tests on monks and the study of dancers mentioned earlier, scientists have done few studies that try to understand what's different about the raw wiring in the brains of talented, especially predictive people. And interestingly, naturally talented people have a hard time identifying why their brains do what they do.

We talked, for instance, to Earle Whitmore, who for two decades has consistently been one of the top real estate agents in the Washington, D.C., area. Her predictive talent is in reading people's needs and intentions when they want to buy or sell a house. She seems to be able to pick up on a situation in a flash and predict what to do to win homeowners and prospective homeowners over. On one occasion we followed her to a meeting with a couple who was considering engaging her to sell their home. When we arrived at the three-story townhouse in the suburbs, the husband greeted Whitmore at the door. In seconds, she assessed the couple's body language and realized the wife was the decision maker. Sure enough, the wife's job was the reason the couple was selling the house and moving to Colorado. Earle made conscious efforts to speak to the husband first, giving him a chance to voice his opinion before turning her attention to the wife. This was to empower everyone, she later told us. "I couldn't take his authority away," she said. Whitmore won the listing.[46]

How does she do it? What goes on inside her head? When we asked, Whitmore responded: "I can't answer it. Except I do what I do. It's probably a combination of acquired skills and innate ability." Whitmore seemed to be naturally good at real estate, even before getting the thousands of hours of practice usually needed to acquire skills in that profession. She grew up in New Jersey, left home at sixteen, and married a West Point graduate at eighteen. She said she went to "many different schools but finished at Vassar," and subsequently worked as a ski instructor, museum director, and art critic and for the Army Corp of Engineers before getting her real estate license. She sold so many houses just working weekends that she quit her regular job and went into the business full time. That was in the 1980s.

Of course, Whitmore has been practicing real estate for so long now that her talent is honed, like Rocca's after a couple decades in show business. But she was good at it before the honing. How? Her brain was wired that way! Which brings us back to where we started.

Same thing was true of Roger Craig, the running back who won three Super Bowls in the 1980s with the San Francisco 49ers. What made Roger Craig so good? Certainly, Craig was physically gifted. But the advantage of those 49ers teams was the intelligence of the offensive players on the field, particularly Craig, quarterback Joe Montana, and receiver Jerry Rice. Craig was known for seeing the play develop and making a quick calculation about where to go, often finding space in a line where others wouldn't. He had this ability as a young man playing in high school, and later at the University of Nebraska. It had to have been largely innate.

So we asked Craig how he processed a football game in progress. He wasn't entirely sure. "I run with my eyes," he said. "I'd study plays and visually see the defense and learn when to cut, and you have to run with your eyes, react, see a hole, and attack it. I made a lot of runs where the play was designed to go one way, and I saw something totally different."[47] And that was pretty much all he could do to explain his two-second advantage. Craig is a smart man. But it seems extremely difficult for those who are especially talented to articulate how their own raw wiring works.

Schmieder, however, was forced to try. It was the only way he could rebuild his life.

"The injury? Oh, this is very bad for me to talk about," Schmieder said to us. "When I came to this country, I never wanted to talk about it. I would not have a job if I mentioned it. I didn't know a single person here. I was penniless. And I was already over thirty years of age, which in our profession is a problem. I could not play as a concert violinist."

He came to the United States with his wife and daughter, Hanna, then just three years old. There was really only one thing he could do to make money: teach violin. But that meant he had to learn whatever it was he already knew. He had come by his talent so quickly and effortlessly that he had never had to really examine how he played the violin so well. "If before this someone asked me how to teach them how I do this or that, I would not have been able to explain because these things were given naturally to me," Schmieder said. "After the injury, I started to think about it, down to the simple things." He broke down notes and sounds into categories and assigned them acoustic and emotional qualities based on how he used to play. He analyzed the mechanical movements he made to produce those different sounds. He had to take apart everything he knew before he could put it back together and teach it to someone else.

"My egocentricity was destroyed by my injury," he says. "Maybe this experience made me understand music and life deeper. Maybe this is why I became a teacher—why some great performers cannot teach."

Before long, Schmieder became one of the most sought-after violin professors in the world—the kind of teacher who would teach future concertmasters. He is a tenured professor of violin at the University of Southern California and in the 1990s founded a global student orchestra, iPalpiti. He eventually returned to performing on violin and conducting.

While teaching, he tries to impart to his students all he has learned about himself—the way to think about notes, the way to draw the bow. He also tries to teach them to learn all of these details, use them to build a mental model, and then forget them. "I think this is where my success lies," he said about his ability to get students to let their intuition do the driving. "We have to bypass the conscious mind and go to the subconscious."

There is something else to Schmieder's story. After coming to the United States, Schmieder and his wife had a son, who is an astounding visual artist—and is autistic. That made Schmieder think about his family history. He recalled that his father could multiply numbers in his head and instantly count a pile of matches dropped on the floor. His father was no doubt to some degree autistic. And, Schmieder concluded, "Maybe that has something to do with me, too." Schmieder is undeniably a fully functioning, engaging adult. But as a child, a degree of autism may have given him a clearer pathway to the "trees" of a musical composition, allowing him to remember and reproduce notes without practicing them much. Yet he had enough high-functioning brain power to chunk information into mental models that could operate efficiently and make predictions—giving him his ability to hear the notes before he actually played them.

Is it possible to build a computer system that is similarly wired for predictive talent? There's a great deal of research to be done to discover more about the wiring of especially talented human individuals, but as that work unfolds, technologists are learning from it to create new kinds of systems that don't work the same way computers have worked in the past.

In the meantime, while trying to learn from the brain's hardware, technologists are also looking closely at how talented brains get programmed.

Talented Software of the Average Brain

Boston mayor Thomas Menino is one of the most successful city executives in recent history. In 2008 a *Boston Globe* poll found he had a 72 percent approval rating, and 54 percent of residents reported having personally *met* the mayor. In 2009 Menino was elected to a fifth term in office. When we talked to him in 2010, he'd been mayor since 1993.

We spent a day with Menino trying to learn what made him so effective.[48] At first, his talent seemed anything but obvious.

Menino arrived on a spring morning to give a brief talk at a conference called the Great Neighborhoods Summit on the campus of the University of Massachusetts. A burly man with thick features and neat hair gone almost white, he was wearing a dark pinstripe suite and light green tie. He told us he has five hundred ties and feels like it's a way for a man to show his individuality. As Menino walked from his car to the conference, his staff gave him some three-by-five cards with prepared talking points, but Menino didn't look at them—he just stuffed them in his jacket pocket. After an introduction by the opening speaker, Menino climbed to the podium and showed why Bostonians sometimes call him Mayor Mumbles. He's famous for his malapropisms, like the time he said at a rally, "Much like a cookie, I predict the Yankee dynasty will crumble and the results will be delicious for Red Sox fans." Menino is certainly no knock-'em-dead orator. At the conference, he seemed more like a favorite neighbor who was uneasily saying a few words to kick off a block party. He started with impromptu remarks, getting a laugh by making fun of Harvard—which went over big with the audience at UMass. Then he pulled out the cards and stiffly read from some of them, meandering from topic to topic before making a couple of good-natured closing jokes and going offstage to sustained applause. The effect was engaging and charming, but not so much that it would make anyone say, "Oh, that's why he's a five-term mayor." In short, he didn't exactly exude charisma.

Menino was never a natural leader or academic genius. Born into an Italian American family in 1942 in Boston's Hyde Park neighborhood, he was a *C* student through high school, thinking maybe he'd become an engineer. He went to Chamberlayne Junior College in Boston, wasn't all that enthusiastic about his education, and graduated with an associate's degree in business administration. For the next few years, he tried to figure out what to do with his life. He tried selling insurance and didn't like it. He worked for Boston's Redevelopment Authority and got interested in local politics. In 1983, at forty-five years of age, he ran for city council. "My wife and I knocked on every door" in the district, Menino says. "There's no magic to this business—it's all hard work."

Menino won the election, served on the council for nine years, stepped in as acting mayor for four months in 1993 when then-mayor Ray Flynn became U.S. ambassador to the Vatican, and in November 1993 won his first mayoral election. He's been unstoppable ever since.

We realized why once we watched Menino work through a day.

In one morning, Menino schmoozed with developers in his office, rode around the city's neighborhoods in a car, appeared at small-time neighborhood events, and chatted up the clientele at the Dry Dock Café in a tired old industrial park he'd been trying to revitalize, while downing a lunch consisting mostly of fried clams. We talked about controversial decisions he'd made to put a skating rink in Boston Commons, merge two hospitals, defy striking firefighters, and defend the city's Muslim population after 9/11. All of those actions had worked to the city's—and Menino's—benefit. Whatever topic came up—a proposed building, a park, a parade, a piece of legislation—Menino seemed to instantly understand all the ramifications it would have on Boston.

"I want to know everything—I can't stay in City Hall," Menino said. "I never do polls or pay attention to them. They're a snapshot of what people thought three days ago, at best. I don't use consultants." He relies on, literally, talking to as many people as he possibly can. One of his longtime staffers, Peggie Gannon, told us, "I'm not sure he has a 'political' mind. It's just hands-on, 24-7. He's not behind a desk. He does things real people do. He's keyed in to little things. He sees the forest and the trees."[49]

Indeed, Tufts University political science professor Jeffrey Berry once analyzed Menino by saying: "He's managed to extend a personal touch that is unusual in a city the size of Boston. In his bumbling, inarticulate way, he does manage to give off a sense of, 'I care.'"[50]

Through the course of the day, we began to understand Menino's talent. Menino kept telling us how on the one hand he does so much personal groundwork but on the other hand doesn't deliberate about decisions or rely on polls or experts. "I don't think a lot about anything—I just do it," he said. He can see the decisions and their impacts in front of him almost instantaneously. He can take in something about Boston and immediately make a highly accurate prediction about its impact. Yet Menino had a hard time explaining how he did this.

Tom Menino was not born with special political talent. He's not brilliant in the manner of a high-IQ Ivy Leaguer. But over the decades he methodically, dutifully learned everything he could about every facet of Boston and chunked that deep knowledge into an efficient mental model that sits inside his brain, instantly recognizes complex political patterns that affect Boston, and analyzes everything that comes before him.

That, in turn, lets him rely on what he calls instinct. He doesn't need pollsters or consultants because his mental model can predict what something means more effectively than any poll or study. Ultimately, Menino has been able to out-maneuver opponents and critics because he can make predictions and calculations about Boston a little better and faster than everyone else—a politician's version of the two-second advantage.

Late in the day, at lunch over those fried clams, we asked Menino if this seemed to describe his talent. "Yes, I think that's it," he said with a wide grin.

This chapter is less about the wiring of the brain and more about its programming—the ways we have developed our brains through practice and experience. It's great to have a predictive advantage because of superior wiring or hardware. That certainly gave Eduard Schmieder and Mo Rocca advantages. If computer hardware could be built to work like an especially talented brain, that would go a long way toward creating predictive ability in computers. But most people don't start out with genius ability, and computer hardware still processes information like a computer, which is to say not at all like a brain—much less like a talented, predictive brain.

So it's encouraging to know that more ordinary brains like Mayor Menino's can be developed or programmed to exhibit a predictive two-second advantage. In the realm of computers, it suggests that we don't necessarily need brainlike hardware to develop some brainlike capabilities using software. For humans, of course, it gives the majority of us hope. In fact, research suggests that most humans who excel in their fields are made, not born with a great genetic advantage.

The leading scientist studying exceptional talent has, in a twist, become something of a superstar himself. Anders Ericsson was born in Sweden to parents who, Ericsson said, "claimed that every healthy person was able to do anything that he set his mind to do."[51] No doubt that attitude had a major impact on him. Ericsson found his way into psychological research. He moved to the United States for his postdoctoral work, wound up as a professor at Florida State University, and edited the 2006 volume titled *The Cambridge Handbook of Expertise and Expert Performance*. The nine-hundred-page book detailed research by more than one hundred scientists all over the world who studied expert performers in fields that included golf, surgery, chess, piano playing, and stock trading. The book caught the attention of the media because it concluded that natural talent is overrated and that years of intense "deliberate practice" could make almost anyone into a star in any field. It seemed ultimate proof for the self-help crowd: you *can* do anything if you put your mind to it.

Ericsson and his colleagues laid out the ten-thousand-hours rule made famous in Malcolm Gladwell's book *Outliers.* That is, almost anyone who intensely practices something for ten thousand hours—years' worth of sustained practice—will become a superior performer in that field. In a 2007 paper, Ericsson and colleagues wrote: "By the age of 20, the most accomplished musicians had spent over 10,000 hours of practice, which is 2,500 and 5,000 hours more than two less accomplished groups of expert musicians or 8,000 hours more than amateur pianists of the same age. The same type of solitary deliberate practice has been found to be closely correlated with the attainment of expert and elite performance in a wide range of domains."[52]

Ericsson defines deliberate practice as the kind of practice you'd expect to see on a stellar performer's résumé: a tennis pro who spent her teen years training at a famous tennis camp; a classical musician who studied under the best teachers and practiced for hours every day; the CEO who went to a top business school and then worked under mentors who challenged him day in and day out. We see those stories in the media all the time. Ericsson cites golf legend Sam Snead, who told him: "People always said I had a natural swing. They thought I wasn't a hard worker. But when I was young, I'd play and practice all day, then practice more at night by my car's headlights. My hands bled. Nobody worked harder at golf than I did."

Such deliberate practice is how a lot of talented people became so accomplished. Through the lens of the two-second advantage and neuroscience research, all that practice generates mountains of data that these people's brains are able to chunk into an efficient mental model. So the tennis pro winds up being able to anticipate her opponent's shots in uncanny ways. The musician can hear notes before they are played. The CEO can foresee the outcome of a decision with amazing accuracy.

How does the idea of deliberate practice play out in Tom Menino? He certainly put in ten thousand hours—no doubt more—to build his mental model of Boston. He is proof that someone does not need to be "talented" in the usual definition of the word to become a master of his profession. Yet Menino built his model in an ad hoc way, on the job. He didn't practice, per se, he just did it. Menino grew up in Boston, worked for the city, helped on some political campaigns, and sat on the city council—but mostly, as he did his work, he soaked up knowledge about Boston and its workings. He pushed himself to get out and see what was happening in the streets and hear what its citizens were talking about, always analyzing and categorizing the information he took in.

Menino's experience turns out to be a variation on Ericsson's deliberate practice—something that Gary Klein (the researcher we mentioned earlier who studies decision making) and his colleague Peter Fadde recently identified as "deliberate performance." While a sports or a music professional can hire a coach and spend isolated time practicing, most of the rest of us can't. An architect or police detective can't hire a coach or attend practice sessions. They might take a class now and then, but that doesn't qualify as the incessant deliberate practice necessary to become predictive and talented under Ericsson's model. Klein and Fadde, however, conclude that the majority of talented people get in their ten thousand hours by constantly challenging themselves on the job, piece by piece building the "tacit knowledge and intuitive expertise" associated with the top performers in a field.[53] "Practice, practice, practice is too simplistic," Klein told us. "In some cases, it's not feasible."

Remember Paula Tallal's research into the reason certain children learn language better than others? When neurons fire together closely in time, they wire that pattern of events and knowledge together. As that pattern is repeated and learned, the chunk becomes more solid, more complex, and faster at accessing the whole pattern. The repetition builds myelin along the pathways of the connections among the neurons that are chunked together, turning the neural networks into superhighways. Deliberate practice and deliberate performance fit with Tallal's research and most other research into the workings of the brain.

It is through deliberate performance that Menino built his mental model and, in fact, continues to build it, which is why he won fourth and fifth terms as mayor. He constantly improved his predictiveness—his two-second advantage. That constant improvement ensures that his predictive advantage is just a little bit better and faster than anyone else's—because in any field, you can bet that competitors are honing their own predictive models and trying to catch up.

While Menino kept adding new information, however, he wasn't creating a bigger and bigger database that he searched every time he needed to make a political decision. That would take far too long—it would be the way Ben Horowitz's "two" managers work, and that's not how an effective, predictive leader works. Remember, as Menino insisted: "I don't think a lot about anything—I just do it." Which means he doesn't go searching his entire database for information or get yet more data from polls or consultants. Instead, he uses the data he collects from constantly canvassing Boston to refine his mental model. The new information gets processed through his mental model of Boston and either confirms and strengthens assumptions built into the model or forces Menino to slightly retune the model. More than likely, after years of so much experience, little can catch Menino's mental model completely by surprise. The model stays agile and efficient, because it is not loaded down by reams of data. The data was used to tune the model and then either got placed into long-term storage (if the information seemed significant enough) or forgotten. The act of forgetting—and knowing what to forget—was as important as the act of storing.

In the end, Menino maintains a finely tuned predictive mental model. When he makes decisions, to him it feels like he's acting on instinct—like he's not thinking at all. But what he's doing is actually better than thinking. He's being a "one" CEO, a Gretzky on ice, a Mo Rocca seeing a bon mot develop in slow motion. Such talent is built on data but not reliant on data.

Which is how businesses need to think about data. Except businesses might be a little more deliberate than Mayor Menino. They might want to build their models in a manner more like the world's greatest pickup artist.

Erik James Horvat-Markovic grew up in Canada as one of those awkward teens who got pushed aside in the halls at school and would have a better chance at winning the lottery than getting a girl to kiss him. He didn't hit puberty until after he turned sixteen, and by then had taken up a hobby that works like antimatter on girls: Horvat-Markovic had become a devotee of the board game *Dungeons and Dragons.* He knew little about being social and nothing about luring the opposite sex and had devoted little brain capacity to figuring it out. He chalked up his bad luck to being shy and being a geek and assumed he could do nothing to change that.

At one point in high school, he read a book called *501 Magic Tricks*. "What was amazing was it wasn't four or five or six different ways of screwing with someone's head and showing the holes in their perception," Horvat-Markovic told an interviewer. "There were 501 ways! All in one book. It was mind-altering to me."[54]

He took up magic and became good at it. His magic tricks helped him become more social. "The turning point for me came in my late teens, when I traveled to Florida to perform a magic show on a cruise ship," he wrote in his book, *The Mystery Method*.[55] Before even starting work on the ship, he made money performing in restaurants, where he was supposed to go up to small groups of people and perform a little magic to entertain them. He went by the name of Mystery. "The experience taught me a series of essential concepts," he wrote. "You shouldn't just walk up to a table of strangers and say, 'Uh, hi. Would you like to see some magic?' The easiest thing for someone to say in that situation is no. So I had to come up with a series of techniques for being cool—that way, they would actually want me to be there." He eventually learned a set of social routines that worked, and then came his revelation: "I could remove the magic from those routines and they would still pack a punch." He could talk to anyone. He could captivate a group. Along the way, Horvat-Markovic physically shot to six feet, five inches tall—and while he didn't have movie star looks, he was becoming an attractive young man. Gaining confidence, he decided he could use his newfound capabilities to pick up women.

Mystery by then had not come close to practicing his new

social skills for ten thousand hours, and he had barely started "deliberate practice" on women. Mystery's mental model didn't really know how women worked, how they reacted to things he might say or do, or why something interested one woman but not another. He started going to nightclubs and talking to women, and he failed miserably, over and over again. And then Mystery decided to get serious and be methodical about his quest. He read books on psychology, which gave him ideas about how to pick up women—ideas he then field-tested at clubs for hours on end, purposely enduring rejections so he could learn from the encounters. He diagrammed a system and wrote out rules for how things worked between men and women. He worked at all this for ten years. "He studied human behavior," observed Mystery's friend, author Neil Strauss. "Until slowly he put it together. The charts. The diagrams. The algorithms. The technical terms. Every day, he wrapped his head around the puzzle of social interaction."[56]

Horvat-Markovic wound up with a set of instructions he called the Mystery Method. "I've backward engineered it from my own success,"[57] Mystery declared. He wrote it all down in his book and created seminars and workshops where he teaches thousands of men how to pick up women. He starred in a reality TV series, *The Pickup Artist*, training hopelessly awkward young men how to talk to women.

It might be easy to think that Mystery is something of a joke. He tends to show up at a club wearing goggles affixed to a floppy hat, bright red lipstick, eyeliner, a pirate jacket, and platform boots. His chosen field isn't economics or engineering or politics but getting women into bed. And yet if you read *The Mystery Method* with a sober eye, you'll find that it's a surprisingly careful psychological and anthropological study.

Ultimately, Mystery's method is about gaining a two-second advantage in a meeting with the opposite sex, and to get there, he preaches deliberate practice. For instance, he pushes newcomers to his method to go to bars and clubs four nights per week, for four hours each night, making three approaches to women per hour. "Within a year, you will have approached more than two thousand women," he said. The result, as Mystery described it in his book, is an exact definition of creating a predictive mental model.

"Patterns emerge over time," Mystery wrote. "Formerly puzzling social behavior comes clearly into focus. Situations and reactions can be easily predicted before they occur." He continued: "It will seem like you're moving at hyperspeed while the world around you slows to a crawl. You don't even have to think about what you are saying. Your mind is free for other tasks, such as planning the next move. It's almost like seeing the future."[58]

Mystery, in his odd little corner of pop psychology, discovered how to purposely, systematically build predictive talent where previously there had been a complete absence of it. He is Ericsson's deliberate practice and Gary Klein's deliberate performance in the flesh, so to speak. Mystery became to picking up women what Gretzky was to hockey. Mystery didn't need to know everything about a particular woman or plan in advance what to say or do. He didn't need a long-range vision—all he needed was that two-second advantage. He could process what was going on in an encounter with a woman in real time and very accurately predict what was about to happen a move ahead. That let him know exactly how the woman in question would react if he made a certain kind of comment or if he touched her arm in a certain way. He had used massive amounts of data collected over time to chunk patterns and build an efficient model in his head. His senses could take in what was happening in front of him, process it through that model, compare it to chunks of memories, and make highly accurate short-term predictions. The result was that Mystery could walk into almost any bar or club and leave with a beautiful woman.

Strange as it might seem, businesses can learn from this. Mystery's process could be a way to create software that lets businesses learn about a market much as Mystery learned about women, and react the way Mystery can react in a bar. The goal is to read the situation, make a highly accurate short-term prediction, and win over a customer. Today some people can do that. The next step is to make machines that can help.

Why pay so much attention to how mental models can be developed? Well, if we're going to learn anything about the brain that's practical for business in the near term, the smart use of software is key for a couple of reasons.

First, as we've seen, hardware doesn't completely account for the predictive nature of talent.

And second, hardwired talent will be too difficult to replicate anytime soon. We're getting a glimpse of that in the early 2010s thanks to what might be called the cat-brain controversy.

In recent years, brain wiring has been a major focus of attention both in neuroscience and in popular books about talent and intelligence. In particular, there's a lot of new research around a substance called myelin, a whitish electrical insulation that sheathes the nerve fibers—or axons—that transmit electrical impulses around the brain. Myelin had been of little interest to scientists in the past, because it didn't seem to do anything important. But in recent years, brain imaging and cellular research have showed that myelin changes during learning. Myelin forms around axons that are excited over and over by experiences, and as myelin covers an axon, the speed of impulse transmission over that axon increases by up to one hundred times compared to a bare axon.

One of the early studies that discovered the connection between myelin and learning involved juggling. Jan Scholz and colleagues at the University of Oxford, in England, used an MRI to scan the brains of forty-eight people—and then the researchers taught half the subjects to juggle. After the juggling lessons kicked in and the juggling half of the group got pretty good at it, Scholz rescanned all forty-eight brains. The myelin needed to connect the areas of the brain necessary for juggling had increased in the group that learned to juggle compared to the group that did not learn to juggle.[59]

One conclusion of the research is that myelin can be built and changed by learning, and as you build myelin you get better at what you're learning. Through repetition, this process moves your brain to the point where something "clicks"—that transition from struggling to learn a skill to doing it intuitively. If you overlay that on the research showing that top performers systematically put in ten thousand hours of practice, you can see where this is going. All that practice seems to build up a whole lot of myelin in all the right places. It's the physical manifestation of deliberate practice.

When you need to tie information together from all around the brain to do something quickly and easily, myelin seems to be a key. Imagine the myelin in Wayne Gretzky's brain—massive clots of myelin around axons that connect his knowledge of hockey with his knowledge of how to move his body to shoot the puck and skate, connecting all of that with information coming in through his eyes and other senses, making for a super-high-speed system within the broader system of his brain. Myelin is the turbocharger in Gretzky's chunks, and it is a reason why Gretzky can react instantly and instinctively in hockey but would struggle to learn something like surgery. The myelin has been built up in one system but not the other.

While entire books describe how to build up myelin and thus become more talented at an activity, the hardware seems to be only a part of the equation. Something else is happening in the brain's software, or programming. Pickup artist Mystery and Mayor Tom Menino don't rely only on fast brain wiring to examine every bit of data they've collected about women and Boston, respectively. They're not consciously thinking through all they've learned in ten thousand hours. A woman at a bar would get bored waiting for Mystery to even get to his second sentence. Instead, Mystery and Menino say they act on instinct and constantly make predictions about what's going to happen. As Mystery said, "It's almost like seeing the future." We know from Grossberg, Hawkins, Olds, and other researchers that the brain of a talented person like Mystery or Menino uses all the information it has gathered to create an efficient, fast-acting model—a little piece of intelligence software that quickly processes sensory input and makes fast, accurate predictions.

Myelin certainly must help the mental model to work faster and quickly get input from the far corners of the brain. But it doesn't explain how the mental model gets programmed. That's still a yawning gap for researchers to explore. As scientists learn about how those models get formed, it will help technologists write programs that mimic the brain's intuition and predictiveness—without having to build computers that operate exactly like a human brain.

This is where the cat-brain controversy comes in.

Various brain research and supercomputing centers around the world have been competing to simulate the workings of the brain. The idea is to start with lower-level species and work up to the human brain, first simulating a small portion of a rat brain, then simulating a whole rat brain, then a cat brain, and so on. The challenge is expected to go on for years if not decades—the task is that difficult. We'll get into more about these efforts later in the book.

In late 2009, one of the challengers—a group led by IBM and including five universities and the Lawrence Berkeley National Laboratory—announced that it had simulated a brain with one billion neurons and ten trillion synapses, which the group said was roughly the equivalent of a cat's cortex.[60] It seemed like the group had made a breakthrough far ahead of the rest of the field.

The announcement, though, made Henry Markram irate. Markram is the director of Switzerland's Center for Neuroscience and Technology, which is running a high-profile brain-simulation project called Blue Brain. He told any media outlet that would listen that IBM's cat brain was extremely simplistic and that a full-scale simulation of a cat brain is a phenomenally overwhelming problem. His reasoning gives you an idea of how overwhelming the problem of detailing brains is: "Neurons contain tens of thousands of proteins that form a network with tens of millions of interactions. These interactions are incredibly complex and will require solving millions of differential equations. They have none of that," Markram barked to a reporter for the Discovery Channel's Web site. "Neurons contain around 20,000 genes that produce products called mRNA, which builds the proteins. The way neurons build proteins and transport them to all the corners of the neuron where they are needed is an even more complex process which also controls what a neuron is, its memories and how it will process information. They have none of that."

Markram kept going: "Synapses are also extremely complex molecular machines that would themselves require thousands of differential equations to simulate just one. They simulated none of this. Then there are glia—ten times more than neurons—and the blood supply, and more and more. These 'points' they simulated and the synapses that they use for communication are literally millions of times simpler than a real cat brain. So they have not even simulated a cat's brain at one millionth of its complexity. It is not even close to an ant's brain."[61]

Whether IBM's team or Markram's team is better at brain simulation is not the point. But the cat-brain controversy showed just how immensely complex it will be to re-create all of the hardware of the human brain in a computer. Markram was arguing that a mere cat brain was far beyond the capabilities of the most advanced computers and the most knowledgeable scientists. It could be a long wait before scientists replicate the human brain's hardware in a machine. In the meantime, however, technologists can be informed by research into the brain's programming—for instance, its ability to chunk information and build predictive mental models—and borrow some of those ideas to change the way computers and business operate.

Joe Lovano came into the Oakland, California, restaurant wearing a purple shirt, black pants, white shoes, and gold-rimmed sunglasses. He looked like either a badly dressed tourist or an appropriately dressed postbop jazz star. Good thing he is the latter.

Lovano is among the most versatile, creative saxophone players in the business. His bands are built to improvise, and he leads them by listening to the players around him, predicting where the music should go, and leading the others there. We met up with him because we wanted to ask how he came by his predictive, intuitive musical capabilities.[62]

The conversation was wildly different from the one with Eduard Schmieder. Where Schmieder felt that his skills and musical knowledge just came to him, Lovano said he had to work hard for them. Born in Cleveland in 1952, his father, Tony "Big T" Lovano, was a working jazz musician playing local clubs. So Lovano probably had the luck of genetics and good brain wiring on his side, but nothing like the wiring that apparently blessed Schmieder. Lovano started playing sax seriously when he was twelve. His father taught him, as did his father's friends and relatives who played jazz. His father's bands often rehearsed at the Lovano house. "By the time I was sixteen, I'd developed enough to play with my father's rhythm section," Joe Lovano said. "That's when my dad's rhythm section didn't cringe when I came in the room."

Joe Lovano just kept working it—chugging toward those ten thousand hours of deliberate practice or deliberate performance. His father started taking him to hours-long jams, which pushed his predictive capabilities to new levels. "I had to be listening fast and learn to hear what a song was doing and realize where it was going," Lovano said. In a classic case of deliberate practice, Lovano attended the famous Berklee College of Music in Boston. At twenty-three, he moved to New York and dove headlong into deliberate performance, joining Woody Herman's big band, the Thundering Herd, and traveling the United States and Europe playing constant club dates. The myelin, no doubt, kept building up in his brain, speeding the connections needed for jazz saxophone. Meanwhile, he kept pulling in data from his musical encounters, using it to refine his mental model of jazz playing, storing important information into quickly accessible chunks while forgetting the rest. Lovano eventually got to the point that came to Schmieder so easily: he didn't have to think when he performed. "It comes from a lot of rehearsing before you get to where you don't think—for me, it's when I got to where I was just emoting and playing from feelings," he told us. "If I start to think too much, it doesn't come out the same."

In the 1990s, when Lovano was in his forties, his career took off, with Grammy nominations, multiple jazz-industry awards, steady gigs and, in 2000, a Grammy win for his album *52nd Street Themes.*

Eduard Schmieder owed much of his talent to brain hardware. Joe Lovano probably has good hardware, but what really made the difference was the great software he built.

Wayne Gretzky seemed to possess both great hardware and software in the extreme, which is why he was perhaps the best ever in his field.

In each case, though, the key was developing that efficient, superfast predictive model that let Lovano, Schmieder, and Gretzky see the future just a little better and faster than most anyone else.

It is an exciting time. What's happening in neuroscience is connecting to what's going on in computer science, and each is learning from the other. This cross-fertilization will result in new technologies that can change the way governments and other organizations do business. Smart leaders will grab the two-second advantage.

PART II
TALENTED SYSTEMS

If It Only Had a Brain

The robots in Rajesh Rao's lab at the University of Washing-ton could be the cousins of the robot from *Lost in Space*, Rosie from *The Jetsons,* and C-3Po. They have square, electronics-laden heads, human-shaped bodies, lights, buttons, and herky-jerky movements. They look like toys but are nothing of the sort. The robots were created to help Rao learn about human brains.

Taking advantage of leaps in computing power and advances in neuroscience, Rao is programming his robots so that they process information a little more like humans than computers. As a robot walks down the halls outside his lab, it anticipates corners and furniture based on previous routes it's taken. If it bumps into something new, that data goes into its hardware brain, and the experience shapes the way the robot walks the corridor next time. In tackling this issue, Rao sits squarely at the intersection of computer science and neuroscience. In addition to Rao's post in UW's computer science department, he also holds adjunct professorships in electrical engineering and bioengineering and is a faculty member in the neurobiology and behavior program. While many neuroscientists use equations and computer simulations to test their theories, Rao is one of the few researchers in the world testing concepts on humanoid robots.

Rao is a slightly built man with neatly combed dark hair and a mustache. He talks fast, and even when he says he has another appointment, he presses on, his passion for his subject overwhelming more prosaic concerns. His office on an upper floor in the new computer science building at UW has a glorious view of Lake Washington. Books pack the shelves and equations fill a whiteboard.

Rao grew up in Hyderabad, in southern India. As a child, he had a Sinclair ZX Spectrum—an early PC that he could attach to his television. The machine hooked him on computers. After he took his SATs, he heard out of the blue from Angelo State University in Texas, which offered him a scholarship, and he left India for the United States. After earning dual bachelor's degrees in computer science and mathematics, he went to the University of Rochester in Rochester, New York, to pursue his master's and doctorate. "I thought I would study the theory of computer science," he says. "But I got interested in the problem of computer vision. I was trying to tackle this problem of recognition, and to do that, you need to have some understanding of how the brain is doing it. So we started looking at the architecture of the brain." [63]

About 1995, Rao started developing his theory of prediction in the brain. He found inspiration in a book, *Large-Scale Neuronal Theories of the Brain*, edited by Christof Koch and Joel L.

Davis (MIT Press, 1994). In particular, a chapter by MacArthur fellow David Mumford pointed to the presence of massive feedback connections in the human neocortex and suggested a possible role for them in generating expectations. Around the same time, Rao became intrigued by the way a number of scientists were studying generative models—internal models of the world that systems (whether computers or brains) develop during the course of their existence. These internal models enable a system to make predictions and constantly compare the predictions to actual events that the system's sensory networks take in. "I began trying to connect some of the ideas suggested by Mumford to the notion of generative models and eventually arrived at a computational theory of how the cortex may employ prediction for perception," Rao says. Our brain perceives things by first predicting what it will perceive, then comparing the prediction and the actual events, and paying attention to elements that are, in essence, a surprise.

By 1999 Rao had published a paper in the prestigious journal *Nature Neuroscience* suggesting that the visual cortex is a hierarchical predictor. "Hierarchical" refers to the brain's levels of perception, from the high-level picture that the brain oversees to the minute details. "The high-level plan would be: I told you to come here," he says. "The hierarchy (at the highest level) was, you had to decide to take your car and take the highway. And then there were all these subpieces, like 'take the key,' and 'turn the ignition.' They're all bound up in this concept of predictions." That fits with other scientists' conclusions about chunking. The more the small details can be bound together and seen as a single pattern, the more the predictions can focus on the higher-level activities.

Rao was also influenced by Hawkins's book, *On Intelligence.*

As Hawkins wrote, "The cortex is an organ of prediction. If we want to understand what intelligence is, what creativity is, how your brain works, and how to build intelligent machines, we must understand the nature of these predictions and how the cortex makes them."[64]

So Rao is trying to understand how the human brain works and using that knowledge to build some of the first intelligent machines—in his case, robots. To get robots to learn as humans do, Rao is starting at the level of human babies. Babies progress from rolling over to crawling to walking along a table to walking independently, and Rao wants to see if his robots can learn to do the same. "When babies are born, they do a lot of body babbling, which is, they flail around, they try to grab things, they do a lot of experimentation," Rao says. "Once they've done that, their brain has a good model of their bodies and how their body interacts with the crib or other people. Once you have a good model of your body, you can make good predictions, you can use it to pick up objects, you can use it to talk with people, and so on. Those are the things that we believe could potentially be implemented in a robot also."

The physicality of the robot is important to Rao. A robot can be outfitted with balance sensors, foot pressure sensors, a gyroscopic system, and other advanced tools that give it more humanlike inputs from its environment. As the robot moves, it can process readings from its devices. "A robot should learn about itself like humans do, in predictive ways. In this case, it's in the millisecond range. It can say, I'm going to move my posture in this way, and then see, Okay, my gyroscopic pressure does this, and my foot sensor feels this," Rao says.

If the robot can chunk the small things, then Rao could

give the robot a higher-level command like, "Go to the kitchen," and the robot could spend its processing power considering where the kitchen is and the best way to get there, rather than how to move its feet and keep its balance. This, Rao says, is the next frontier in artificial intelligence—or perhaps the first frontier in predictive, talented machines.

Rao also believes that the brain not only makes predictions but also computes the uncertainty associated with its predictions. "If you have learned a task really well, your predictions will be accurate and the uncertainty associated will be low," Rao says. "On the other hand, if you are a novice or if you are in an unfamiliar environment, your predictions will have a high uncertainty associated with them. We believe that the brain keeps track of the uncertainty associated with its predictions and actively uses this information as it makes decisions and executes actions." Interestingly, that same concept is a major part of the software in IBM's Watson computer that in 2011 won against humans on *Jeopardy!*

Ideally, Rao says, once scientists understand how the human brain works, robots can be given the task of solving problems in a human fashion. Eventually, if a robot could get the equivalent of ten thousand hours of deliberate practice, it might develop talent. In the shorter run, though, Rao is figuring out how to get a machine to pull in massive amounts of information, use it to build a model, chunk the rest, and then be able to make short-term predictions about the world it encounters.

Rao's early insights from the crossroads of neuroscience and computer science are informing ideas about computing and how computers might work if they worked even a little more like the human brain. Those insights couldn't come at a better time for companies and society.

By 2010 the "data deluge" had become a hot topic of conversation among technology and business types. If you think back to 1990, data found its way into computer systems in relatively few ways: credit card and bank transactions, back-office systems that kept track of payroll and inventory, supermarket checkout scans, land-line phone traffic, a smattering of personal computers on Prodigy, and not much else. Starting in the late 1990s, the Internet and consumer Web sites changed that in a big way. Today, nearly two billion people worldwide are on the Web, and they're shopping, writing Twitter updates, posting videos, listening to music, and doing work. Mobile phones have taken off too, and in 2011 there were nearly five billion mobile phone subscriptions globally. All that activity, including location-based services that rely on global positioning satellites, is flowing into computer databases.

On top of that, the physical world is increasingly packed with digital sensors, all feeding data into storage somewhere. Sensors monitor water flow in rivers, traffic on city roads, what TV shows are getting watched, the sugar level in vineyard grapes, shipments of pencils, the movement of everything from freight train cars to whales in the oceans—and much more. On a cross-country flight, the sensors in a Boeing 737 generate about 240 terabytes of data.[65] If all the data from all the jets in operation were saved in computer storage systems, it would overwhelm every airline-company data center on the planet.

The data deluge keeps gaining speed as people increasingly do work and live life on the Web and sensors move deeper into the physical world. The world's storehouses of digital information contained 281 exabytes of data in 2007 and will hold six times as much by 2011, an increase to 1.8 zettabytes. These numbers are unimaginable. (A typical PC user can get his or her head around a gigabyte. A terabyte is a thousand gigabytes; a petabyte is a thousand terabytes; an exabyte is a thousand petabytes, and a zettabyte is a thousand exabytes.) Gartner Group predicts that enterprise data in all forms will grow 650 percent over the next five years, while research firm IDC claims the entire world's data doubles every eighteen months.[66]

Theoretically, all that data should be a godsend. The more we know, the better we should be able to do things. If a company knows more about its customers, it should be able to better please those customers. If a city knows more about its traffic, it should manage it better and be smarter about the roads it builds. If an individual knows more about what she eats and how she burns calories, she should be able to find her optimal way to keep fit. As data gets deeper and more precise, everything it touches should become more optimized, productive, and efficient.

But we've hit a snag. There's so much data, today's systems are increasingly having a hard time finding and processing the right data to get the best answers. A 2010 study by Avanade found that 30 percent of business leaders say they can't get the data they need at the speed they need it, and 61 percent want faster access to data. Not many people want *less* data—in fact, one in three executives wants *more.*[67] And yet a study by the Massachusetts Institute of Technology in 2010 reported that 60 percent of executives say they "have more information than [they] can effectively use." Still, companies and governments continue to add sensors to physical things and collect ever more data from electronic activity. The influx of data will grow unabated for the foreseeable future.

At the same time, the ways to glean insight from all that data have changed slowly. Computers operate pretty much the same way they've operated for sixty years, just exponentially faster. Database, simulation, and analytic software programs churn numbers through algorithms one step at a time, in sequence. And even though the machines do each calculation in nanoseconds, as the data piles up, pushing that data through sequential algorithms is like forcing an ocean through a straw. The solution, so far, has been a brute-force approach: add more computer processing and parallel processing—i.e., add more straws. But even that can't keep up. Crunching through all the available data to gain insight will take longer and longer if computing models stay pretty much the same.

That's a big problem, because the time to react is getting shorter and shorter.

As discussed earlier, for most of the twentieth century companies operated in a mode that might be called Enterprise 1.0. Those were the simple days when computers sat in back rooms, touched by only a few people. The time to react to information seemed fast at the time, but from today's perspective it seems glacial.

Modern computing systems brought us into the world of Enterprise 2.0, dramatically reducing the time to react. Depending on how much data had to be sorted through, an answer from a computer might come in seconds, hours, or days. A bank executive could find out what was happening at a branch much more quickly—and the bank's competitors could move equally fast. The pace of business sped up.

Now, in the 2010s, we're in a third stage—Enterprise 3.0. Data is everywhere, all the time. A bank customer might be at a branch, at an ATM, on a Web site, or tapping in via mobile phone. If a customer isn't happy with, say, a money-market rate, she can instantly see competing offers online. If a bank executive has to ask a question of data to gain insight or take action, the answers will come too late. The pace is so fast, systems now have to respond to *events*, not questions. They have to be able to know when something happens and react right then, automatically. Systems have to anticipate and predict. A little bit of the right information a little ahead of time is often more valuable than a boatload of information that lands too late. If the bank customer is online, the system needs to know by her activity that she's about to be interested in a good money-market rate and offer a deal that beats competitors' *before* she looks elsewhere. There is no time for a manager to ask a question of a database. The system has to perceive what's happening and react not just instantly but ahead of time—yes, like Gretzky.

In 2010 the San Francisco Giants won baseball's World Series. One advantage the team had over opponents might have come from technology provided by Sportvision. This is the company that created the yellow virtual first-down line you see on NFL television broadcasts. Sportvision also makes a product called FIELDf/x. Two computer-vision cameras on towers high above the Giants' AT&T Park track the players' movements during games, capturing more than 600,000 location points in a game. The information flows into a database, where a computer can crunch it together with data from past games and analyze which players have the best range on defense, which players take the most efficient path to the ball, and so on. Crunching a season's worth of data can help the Giants' management instruct players on how to play better defense, choose which players to put at which positions, and better decide which players are valuable enough to keep on the team.[68]

In Switzerland, Cablecom is the nation's largest broadband cable TV and Internet provider. It recently unleashed database analytics software on the issue of customer defections. By analyzing customer behavior and calls to customer service, Cablecom found that while defections of new customers peak in the thirteenth month, many customers start making the decision in the ninth month. Cablecom then began offering promotional deals in the seventh month, and it worked, reducing defections by one-fifth.

Both are examples of mining databases and coming up with useful insights. Still, both the Giants and Cablecom are asking questions of the data, crunching numbers, and getting answers after the fact. It's obviously helpful, but maybe not as helpful as it should be in the era of Enterprise 3.0. The Giants' system could go a step further if it had something of the real-time gut instincts of a veteran baseball coach and in the middle of a game—based on events happening right then—could predict that substituting a different shortstop with a different set of capabilities would help shut down the opponent's offense. Cablecom could save a lot of money if it didn't have to blanket every new customer with a discount in the seventh month, but instead could notice which individual customers were growing unhappy and only offer those customers the promotion, just as they were becoming disgruntled.

In the 2010s, a couple of technology trends are converging to help move computing systems toward Enterprise 3.0. One is real-time computing; another is predictive analytics.

Real-time computing means looking at data as it comes in and gaining some understanding about it right then. A simple version of this will help champion swimmers improve during practice. It's a system from Avidasports of Harper Woods, Michigan. A coach using the system attaches sensors to the head, wrists, and ankles of a swimmer. The sensors wirelessly send a stream of information to a laptop measuring the swimmer's pace, stroke count, stroke tempo, turn time, kick tempo, and several other metrics. As the data comes in, it's compared instantly—during the swim—with information from the swimmer's past practices, so the coach can see if some part of the swimmer's performance is off. The coach can then speak to the swimmer through a wireless earpiece and tell the swimmer to adjust her kick tempo or stroke count. The data is helpful in the moment.[69]

A project at the University of Ontario Institute of Technology in Oshawa, run by health informatics professor Carolyn McGregor, helps premature babies by adding predictive analytics to real-time computing. The babies are hooked to sensors that track seven streams of data, including respiration, heart rate, and blood pressure. The electrocardiogram alone records one thousand readings per second. By matching the incoming data streams with known patterns, the system can watch for the beginnings of a potentially fatal infection before the symptoms would otherwise be seen by a physician. By getting a little ahead of the situation, doctors can have a better chance of treating the infection.[70]

McGregor's project is a small example of the technology becoming a little more predictive. Projects like it are popping up all over the world—some limited, some ambitious. These are a step toward brainlike talent—toward the two-second advantage that could make a hospital or company or government agency anticipate events like a talented human.

India's cell phone boom has been a cacophonous, crazy miracle of the twenty-first century. A country that had little communications infrastructure just twenty years earlier unleashed a whirlwind of demand as cell phone towers invaded the landscape. By the spring of 2010, 550 million people in India had cell phones—from virtually zero in the 1990s—and the various providers were together signing up twenty million new customers a month. (The largest U.S. cell phone provider, Verizon Wireless, had about ninety-three million customers in 2010. India was adding a Verizon Wireless–size subscriber base every five months.)

The cell phone business in India works differently from how it works in the United States. Most subscribers don't sign up for long-term contracts, and any phone can be used on any of a dozen providers' networks. Competition among the providers is intense, because any customer can switch at any time. A moment of dissatisfaction, and—boom—the customer is gone.

Among the cell phone providers, Reliance Communications is the biggest. Every day, Reliance gets three million "recharges"—a customer adding a bucket of minutes or text messages to his or her phone. That's 150 to 200 recharge events a second. The company has been adding 150,000 customers a day—while at the same time losing tens of thousands to competitors. All this activity was generating vast amounts of data, yet the pace of activity left no time to analyze it. Reliance wanted to tackle the problem of customer churn and find a way to keep customers from bolting to competitors. Doing it the usual way—plowing through past data to find trends and insights—just wasn't going to cut it in this frenetic situation. This is what led Reliance to consider trying to act like an organism instead of a company.

"How do you develop the subconscious of an organization?" said Sumit Chowdhury, the chief information officer of Reliance Communications in the late 2000s before he departed for IBM. "There's no time to take decisions deep into the brain—no time to take it to the data warehouse. It has to become a reflex action. We're developing the gut."[71]

Reliance's gut is gearing up to stop churn. As Chowdhury explained, Reliance wanted to be able to predict the moment when a customer began to get frustrated and considered jumping to another carrier, so that Reliance could in that instance offer a discount that made the customer decide not to leave. But Reliance didn't want to give discounts to every customer who reached a certain point in time—as Switzerland's Cablecom did—because then it would be losing money on deals it gave to people who were not going to churn. Reliance's research found that a specific series of events—like a certain number of dropped calls in a certain amount of time for a certain kind of customer—often meant that the customer was going to bolt. An offer of a promotion precisely at that moment would usually keep the customer. How could Reliance watch for such specific sequences among 120 million customers across a vast nation and instantly react to them?

"We couldn't take it to the database—that would take twenty-four hours," Chowdhury told us when he was still at Reliance. "We had to be able to make a decision like *that*, without reprocessing all of our learning. So we take a small slice of information and a profile of the customer, set up a promotion, prime up the customer's history, and wait for the events. It primes up just the right amount of information in real time."

The capability quickly helped reduce churn at Reliance. But Chowdhury was not satisfied. He believed such systems had to operate with Gretzky-like talent, processing events through an efficient model and making instant, accurate predictions. The Reliance system was still relying on precise rules that had to be written by programmers. It didn't learn about situations and adjust. It couldn't constantly watch customers, use that information to build better models of their behavior, make predictions, test the predictions against what really happened, and refine the model even more. A system like that, Chowdhury explained, "might not be as precise, but it would be statistically good enough. Maybe it would be wrong five percent of the time, but that's okay. It would be more like making judgment calls." In other words, it would be more human in its decision making. The speed of being able to make those judgment calls would more than make up for the times those decisions turned out to be wrong. If the system could instantly predict which customers were about to leave and be right most of the time—and right more often than competitors—then Reliance would win.

In other words, Reliance is chasing the two-second advantage. If a company has just a little bit of the right information a couple of seconds in advance, it's more valuable than all the information in the world months after the fact. The data deluge and the ever-shorter reaction times are putting pressure on companies for a new kind of solution. The arrival of superfast computers that can process streams of data in real time, combined with pioneering work with predictive analytics, is seeding new ideas for predictive technology and new ways of thinking about how organizations can behave more like organisms. Insights from data, which used to be available only after the fact, can now arrive instantaneously and continuously. The next step is for those insights to arrive beforehand. If the computational neuroscientists are right, the inner workings that make talented people exceptionally successful can make companies exceptionally successful.

Little by little, it's beginning to happen. The retail industry, for one, is actively embracing two-second-advantage technology.

Retailers have been making predictions about future purchases since the dawn of commerce. When a grocery store puts out shelves of candy during the weeks leading up to Halloween, or an office-supply store sets up a display of school supplies in late August, they're making predictions about demand. Obvious seasonal promotions have been overshadowed in recent decades by more sophisticated data mining, in which a retailer analyzes all sales transactions and figures out how to cross-market products that sell well together. If you know that customers who buy tortilla chips also tend to buy salsa, you can put those two items next to each other on the shelf. Other algorithms look at demographics or weather patterns. Wal-Mart, for instance, noticed that certain products—including, oddly, Pop-Tarts—sell like crazy just before a hurricane hits. So Wal-Mart knows to rush those products to stores in a hurricane's path.

The ability to make predictions about specific shoppers, though, is a new frontier.

Rent-A-Center is beginning to go down that path. The Plano, Texas–based company operates 3,000 stores that rent everything from TVs to lamps to beds to washing machines. In the past, the company's salespeople didn't know much about any individual who came in a store. "We knew who our customers are, but what we lacked was insight into the products they prefer, the patterns of their rental behaviors, the peculiarities of their demographics, and the changes in their lives that would drive future rentals," said John Gideon, Rent-A-Center's senior director of data management.[72] A customer might rent a seven-piece living room package, and then come back two weeks later for a lamp. If the customer were to engage with a different salesperson, that salesperson wouldn't have any idea that the customer might be open to renting a big-screen TV to pair with his just-rented living room.

Rent-A-Center has been implementing a system that captures information about a customer and starts to build a model of that person. Some of that data is simply what the customer rents and basics such as the customer's age and location. But salespeople are also supposed to enter things they learn about each customer in conversation—a color preference, ages of kids, size of house. As data about customers gets built up, the system can analyze general data about how customers behave and apply it to a particular customer who walks in the door—and make a prediction. It could see that, say, Joe and Rita's past rentals add up to a strong possibility they'll want to rent baby furniture. The system could then flash that on the salesperson's screen, and the salesperson could make the suggestion. Sam's Club figured out how to make predictions so accurate, it became concerned customers would feel that Sam's was creepy.

The Sam's Club system, called eValues, was launched in 2009. It was developed for Sam's by FICO, the company formerly known as Fair Isaac, which is known for analyzing individual credit histories and assigning people a score that summarizes their creditworthiness. Shoppers who enroll in the store's "plus"-

level memberships get a card that entitles them to special discounts tailored to them based on products they've bought in the past. Shoppers can find out their discounts via e-mail, by scanning the card at an in-store kiosk, or by logging in to the Sam's Club site. There's also a mobile app. The system doesn't involve any physical coupons to clip or cut—the price reductions get applied at the register.[73,74]

Behind the scenes, Sam's software first culls billions of historical transactions, sorted by individual shoppers. This way, Sam's can build software models of each customer from the intelligence in the database. The system is basically chunking individual customer profiles. Then the system can leave the database behind and react to the shopper's events. The system looks at more than just previous purchases to find patterns. It also keeps track of the predicted date when the customer might need to stock up on something he's bought in the past. For example, Sam's factors in how long it takes a typical family of four to go through a ten-pack of toothpaste, so that a few weeks before the last tube runs out, the algorithm offers that shopper a discount on toothpaste. Sam's found that seemingly random purchases can be highly predictive of life-changing events like the birth of a child or a divorce, and the system could know if a certain person was very likely to buy a big-screen TV in the next two months—and make an offer a little bit before that shopper even started looking for a big-screen TV. To help make sure customers didn't feel like Sam's was peering too deeply into their lives, Sam's made the decision to tell its premium customers exactly what it was doing, pitching the predictive offers as a benefit of premium membership.

Customers responded positively. While traditional coupons have a redemption rate of around 1 percent, Sam's Club eValues promotions are redeemed at rates of 20 percent to 30 percent. That will probably improve over time, as the system incorporates new sales data to enhance the accuracy of its predictions.

Retail is one sector turning its attention to the two-second advantage in a big way. But as we'll see in the next chapter, innovative ideas about building brainlike predictive talent into machines are popping up in everything from casinos to power companies to vineyards. Along the way, they're changing the way executives think about organizations. A two-second advantage might not sound like a big advantage, but entities that embrace it can become Gretzkys, moving to where the puck is going to be, instead of where it's been.

Talented Technology and Talented Enterprises

Jose Cordero spent nearly a decade trying to bring the two-second advantage to police work, in a town that desperately needed it.

East Orange, New Jersey, sits next to Newark. It's a ten-minute train ride from New York, and then a short car trip through worn neighborhoods and blocks of tattered storefronts. In the early 2000s, East Orange had one of the highest crime rates in the nation. Gang violence and gunshots were a part of life, daring innocent citizens to venture from their homes. More than 1,900 cars were stolen in East Orange in 2003—an average of more than five a day in a town of about seventy thousand residents. The underfunded police force of three hundred couldn't keep up. It was losing the city.[75]

By 2010, the story had changed dramatically, and a big reason was Cordero, the police director. With his boldly styled suits and mustache, he looks the part of a 1940s jazz bandleader, except for the BlackBerry that buzzes constantly in his belt holster. On a visit to his office in the summer of 2010, Cordero showed us an electronic dashboard displaying up-to-the-second police activity spread across a wide computer screen. Two Dell PCs hummed under the desk, feeding the screen statistics, images, and analysis. This system, the Law Enforcement Electronic Dashboard (LEED), helped Cordero and his team accomplish the improbable. From 2003 to 2009, overall crime incidents in East Orange dropped 71 percent. Murders, robberies, and car thefts dropped off a cliff. Gangs, fed up with the suddenly intense attention they got from East Orange cops, apparently moved their operations elsewhere. Cordero's system was so effective, police executives from Brazil, Turkey, and other parts of the world kept visiting to learn about it. But while Cordero talks up his force's predictive technology, he emphasizes that the technology is only part of it. The technology allowed East Orange to think completely differently about what police do.[76]

Cordero went into policing with a background in technology. He graduated from the New York Institute of Technology, then joined the New York Police Department and stayed for twenty-one years. Along the way, he worked with New York's breakthrough crime database, CompStat, created under New York police commissioner Bill Bratton in the 1990s. Insights from CompStat were important to New York's success in driving down crime rates in the 1990s. By the time Cordero was hired as East Orange's police director in July 2004, he felt there was room for a next generation of crime technology—a system that went beyond data mining and moved into predictiveness.

The East Orange police didn't have much money to spend, so Cordero started the process by writing some database software at home to find patterns in past crime data. From there, his team kept adding pieces. GPS devices tracked patrol cars and placed them on a map. The department bought gunshot detectors from ShotSpotter that could pick up the sound of a firearm going off and report the location. Police installed video surveillance cameras and later smart cameras from DigiSensory Technologies that can independently watch for patterns (like a group of people gathering where they shouldn't be gathering) and alert police. The department added a Web site that let citizens anonymously point out suspicious activity, creating yet another data stream. At each step, Cordero sought to pull all the input from all the technology into a single system that could analyze and cross-reference the streams. That way, the detection of a gunshot could be matched to images coming in from a camera, and then the police could use GPS to identify the nearest police car to call to that location. All in all, Cordero figured he spent about $1.4 million on the system—almost nothing compared to the money big companies spend on IT.

As the streams of data developed into Cordero's dashboard, Cordero wanted the system to learn from all that data and begin to predict what was about to happen. He wasn't looking for broad conceptual predictions weeks or days in advance but very specific predictions minutes or even seconds in advance. Longer-term predictions drawn from database analysis can give a police department some ideas about when and where crimes might occur. Shorter-term predictions, driven by actual events and a deep knowledge of crime patterns, are more likely to be highly accurate, telling officers when and where a crime is going to occur. If a prediction is just far enough ahead of a crime, police can actually stop the crime from happening.

So the department tuned the system "to identify series of

actions that by themselves might mean nothing, but a certain sequence means that a crime is likely to occur," Cordero said. One example: smart cameras might notice that a person is walking after midnight on a particular street where a lot of robberies have occurred around that time of night. That doesn't necessarily send alarms through the system. But then the cameras might notice that a car approaches and slows near the person. "Now I want to know about it," Cordero said. The system alerts an officer sitting in headquarters at a screen, watching the data dashboard, and it flicks on the nearest surveillance camera. The officer can see the image and instantly recognize a problem. The system finds the nearest squad car—which is usually about a minute away from any spot in the city. By turning on the car's siren, the patrolling officers can have a nearly immediate impact on the potential crime scene. The robbers bolt instead of carrying out the crime.

As East Orange's system gave the police a two-second advantage, the technology allowed Cordero to push a new kind of philosophy on the department. The measure of success of a good police department isn't how many criminals it catches, Cordero preached—it's the absence of crime. "The real work of police is to have an impact so they slap on cuffs fewer times, not more," he said.

And Cordero took the philosophy one step further: he wanted East Orange police to use the predictive technology to get inside the heads of criminals and gang members and wayward teenagers and spook them. The system could recognize a pattern and alert an officer to walk down a block where a car thief was just about to steal a car. Or the system could send a squad car to cruise by a vacant lot where a drug deal was about to occur. The bad guys would feel like the police were specifically watching *them*. When police made arrests thanks to the technology, officers would make sure to show the suspects how the technology had spotted them, knowing those suspects would tell their friends. Cordero wanted word to get out that the police had a two-second advantage. He wanted every criminal in town to think that at any given moment, the police might know exactly what he or she might be planning to do. The technology allowed the East Orange Police Department to be reengineered to focus on driving potential criminals crazy—which, in effect, probably drove many of them out of town.

"It was hard to sell to officers at first," Cordero said. "But once we got it going, they began to believe in it." The community, too, bought in. Crime had been so bad, the citizens were willing to risk some privacy concerns for the benefits of a safer town. The department has also tried to build privacy safeguards into the system.

East Orange is by no means the only police department diving into predictive policing. Memphis, Los Angeles, and other major cities are investing in predictive systems, and the issue of how well the systems work has become a hot topic in criminal-justice circles. Universities are playing important roles, studying the anthropology of crime and using it to help build software models that can provide a two-second advantage. "The naysayers want you to believe that humans are too complex and too random—that this sort of math can't be done," Jeff Brantingham, an anthropologist at the University of California at Los Angeles, told the *Los Angeles Times*. "But humans are not nearly as random as we think. In a sense, crime is just a physical process, and if you can explain how offenders move and how they mix with their victims, you can understand an incredible amount."[77]

East Orange, so far, seems to show that the predictive model works. But the city has also proved a couple of other key points. One is the value of fast, short-term, highly probable, Gretzky-like predictions about crimes that are likely to happen in just a few minutes, compared to relying on data mining to predict patterns weeks or months in advance. The other point is that the two-second advantage is not only about technology—it's an entire organizational mind-set. East Orange police became much more effective when they stopped thinking about criminals as people to catch and instead thought of them as people to influence.

The East Orange system is far from perfected. Two-second-advantage technology is new, and any company or government entity moving into this kind of technology is a pioneer. Real-time computing and predictive analytics go a long way toward anticipating outcomes based on events, but the next iteration—to brainlike talent—is still in a primordial state, emerging and evolving right now. As we'll see, there are a number of enterprises in a wide variety of industries starting to deploy early two-second-advantage systems. As in East Orange, those systems have come about because leaders thought about data and the organization in a new way.

The concept of predictive organizations has been evolving for a decade. In 2002 Kemal Delic and Umeshwar Dayal—two Hewlett-Packard scientists—published an influential treatise, "The Rise of the Intelligent Enterprise."[78] They argued that organizations of the future will increasingly resemble natural-born, organic systems in their ability to sense, learn, and evolve. Delic and Dayal made the case that today's enterprises exist in a fast-changing, competitive environment in which the basic objective is to thrive and endure—much the same goal as a sentient being. Ultimately, the most successful enterprises will develop a heightened sense of awareness and an ability to learn and adapt—and eventually they'll develop brainlike predictiveness and talent. An intelligent enterprise will respond like a living thing, not a giant bureaucracy.

Ideas about artificial intelligence (AI) in machines have been around a very long time. In 1950 English mathematician Alan Turing published a landmark paper saying it's possible to create machines with true intelligence. He described the now-famous Turing test: if a machine could carry on a conversation with a human being—via typed words or any other medium—so that the human wouldn't know it was a machine on the other end, then the machine would be "intelligent." As of this writing, no machine has passed the test, but beating the Turing test remains a goal in AI.

Through the decades, fascination with and investment in AI has waxed and waned. In 1970 Marvin Minsky, a famous MIT scientist, told *Life* magazine, "In from three to eight years we will have a machine with the general intelligence of an average human being. I mean a machine that will be able to read Shakespeare, grease a car, play office politics, tell a joke, have a fight. At that point the machine will begin to educate itself with fantastic speed. In a few months it will be at genius level and a few months after that its powers will be incalculable." Of course, that has not come true, and nearly four decades later, Minsky told an interviewer: "Oh, that *Life* quote was made up. You can tell it's a joke." Most people didn't take it as a joke in the 1970s.[79]

By the 1980s, AI had become more focused on "expert systems"—narrow computerized functions that could have humanlike knowledge. The systems were often programmed by trying to take the knowledge of experts and break it down into "if this, then that" rules that could be calculated at high speeds by computers. It was thought that expert systems might capture the knowledge of, for instance, a master brewer at a brewery or a receptionist answering calls. Expert systems haunt us today in the form of the automated voice-response systems we get when we call customer-service help lines. Those call-center systems have made millions of people hate AI.

But the path toward expert systems led to the successful use of AI in very specific ways. The technology behind the autopilot systems on airliners is AI. So is the intelligence that tells a Roomba vacuum cleaner how to move around a room. The antilock brakes on a car are essentially a tiny artificial intelligence processor gauging pressure on the brakes and rotation of the tires and taking over the braking system for a moment. In many ways, AI has melted into everyday technology that aids and augments humans but has no pretense of being human in any way.

At a higher level, one of the most famous expert systems is Deep Blue, the chess-playing IBM computer that beat world champion Garry Kasparov in a tournament-style match in 1997. Deep Blue was programmed with expert chess knowledge, but the key to its success was its brute-force approach to the problem. The massive computer—the 259th most powerful supercomputer at the time—could consider two hundred million possible moves per second. Deep Blue worked because chess can be narrowly defined, and the machine could focus all its computing power on a relatively small set of rules. Deep Blue wasn't talented. It couldn't learn and then apply that learning, or be creative. However, it could do something interesting: in a match, Deep Blue evaluated the probability of what might happen six or eight moves ahead, based on whatever move Kasparov had just made. The machine started to be predictive, though it took a supercomputer and complex programming to do it. Deep Blue showed that within a narrowly defined task, a machine could take in events as they happened and make short-term, highly accurate predictions in much the manner of a talented person. In fact, Deep Blue was good enough at this to gain a two-second advantage over the world's most talented chess master.

The lessons of artificial intelligence suggest that predictive talent will first emerge in systems that are narrowly defined. It will then spread and take on more complex tasks. As predictive talent develops in systems, the talented systems will change the way leaders think about information and organizations. We'll see the emergence of entire predictive enterprises that respond to events with the two-second advantage of talented humans—much as Delic and Dayal foresaw.

And, of course, all of this will happen in a messy, chaotic, stop-and-start manner—because that's the way these shifts always happen.

Predictive, talented systems will be built around the idea that a little bit of the right information just ahead of time can be more valuable than a boatload of information later. We've run into a number of companies from a range of industries that are working on that.

THE TALENTED VINEYARD

A million variables go into the making of a bottle of wine, but nothing is more influential than the amount of water in the winery's grapevines. Yet so much of what most vintners do regarding the management and distribution of water is based on guesswork or on a database of past weather and water patterns that may—or may not—predict future patterns.

Fruition Sciences, founded in 2007 and based in California and France, puts sensors directly on grapevines. The sensors can tell a vineyard manager exactly when a plant is thirsty and how much water it requires. The data from the sensors feeds software that can predict just how sweet a winemaker's grapes will be at harvest—a crucial factor in determining how the wine will taste and what its alcohol content will be. The system reacts to events such as weather and water levels in the vines and gives vintners a little bit of predictive insight so they can adjust and harvest grapes that will make the best wine. Fruition is, in a way, trying to reproduce the instinctive talent of a great vintner, while adding an ability to measure water levels inside the vines. Even a great vintner would have a hard time doing that.

Fruition's cofounders bring together backgrounds in wine and technology. Thibaut Scholasch was a winemaker who saw the need for a better way to manage water in the vineyard. He earned master's degrees in viticulture and oenology in 1997 and a master's in winemaking in 1998, traveling the world to learn his craft. "Everywhere I found the same problem: irrigation," Scholasch told us. "Irrigation was a big issue in Tasmania, in Victoria, in Argentina, in Chile. In the U.S., I got hired by Robert Mondavi. Instead of becoming a winemaker, I transitioned to working on how irrigation processes could be managed in winemaking."[80]

Sébastien Payen started out as a mechanical engineer dabbling in biotechnology and the use of sensors. He earned an engineering degree in France, served a year in the French navy as a deck officer on a mine hunter, and then earned master's and doctorate degrees in mechanical engineering at the University of California at Berkeley. As part of his work, he earned patents for his design of micro-biosensors that can detect changes in pH using plastics and polymers. A French telecommunications company and the state of California gave him a grant to study how to use sensors in vineyards.

While vineyards have employed sensors, the sensors haven't been monitoring the plants themselves. They are embedded in the soil and set up to collect data on the climate. "Wineries are totally filled with sensors at every level of the process: temperature, yeast count, relative humidity in the barrel aging, oxygen levels, all of that is very well controlled," Scholasch said. "It's as if the technology stopped at the gates of the winery and didn't expand into the vineyards."

He added: "Due to climate change and global warming, people were noticing that the alcohol level was getting higher. Because the fruit was losing water, the sugar became more concentrated, and "then there is more alcohol in the final wine." To compensate, winemakers were irrigating early in the season, but "any kind of plant that gets an abundance of water supply early in the season doesn't feel a need to start ripening," Scholasch said, which made the fruit too pungent later on. The lack of ripening, he says, led to "green flavors in the fruit at harvest time."

They started selling their system in 2008, to five high-end vineyards that had helped sponsor Scholasch's research. (The vineyards: Harlan Estate, Ridge Vineyards, Dana Estates, Ovid Vineyards, and Spottswoode Estates Vineyard and Winery.) The next year, the number of vineyards doubled, and it doubled again the next year as the word got around.

The next step for the private company is to use the data it's collecting to make the model more predictive. The company is starting with increasingly accurate and hyperlocal weather forecasts, taking local data and running algorithms that can spit out more precise predictions about what water levels vintners need to maintain to get a precise balance of sugars in their grapes. Having that kind of information just a little ahead of time will give the vintners enough time to act and will be more accurate than long-term predictions based on past data. A really good predictive model is probably "a few years down the road," Scholasch told us. But it's already having an impact on the way wineries manage their grapes and water.

THE END OF TRAFFIC JAMS

Around most big cities, commuters can look at an online map or peek at traffic cameras before heading out of the house in the morning and get a sense of what traffic is like right then. Smart-phone apps with GPS show traffic on roads as you're driving. Yet none of that really solves the traffic problem. It doesn't help drivers know, for instance, if a road twenty miles away will be backed up once they get there, or if an alternate route would take a longer or shorter amount of time. Traffic systems at best only understand what's happening right now or can guess at what traffic will be like at a particular time on a certain day based on past data. The systems have a much harder time reacting to real-time events by making highly accurate, truly useful predictions about what's going to happen in just a little while.

Bryan Mistele was at Microsoft when he first realized that traffic technology could become predictive. He'd joined the company in 1994 after getting his BS in computer engineering from the University of Michigan, earning an MBA from Harvard, and then working for Ford. Mistele had started MSN's personal finance site, then developed a real estate site that Microsoft sold to Fannie Mae. He moved on to develop the in-car computing system that became Ford SYNC. "That's when I started seeing what was happening with dynamic data delivery to cars," Mistele told us.[81]

Microsoft Research—the Microsoft R&D lab with a six-billion-dollar annual budget—had spent ten million dollars developing traffic-prediction technology that Microsoft didn't know what to do with. So Mistele got backing from Microsoft and a couple of venture capitalists, licensed the company's traffic technology, and in 2004 launched INRIX. The company's technology uses past data to build a model of traffic in a given city. Mistele's team also fills the database with traffic-effecting ingredients, including school schedules, construction plans, scheduled events like pro sports games or festivals, and weather data. All of that is supposed to help the system understand that on a school day when there's an afternoon baseball game and a thunderstorm warning, certain roads are likely to be a nightmare.

But that's a generalized guess, not a two-seconds-ahead accurate prediction. INRIX needed real-time sensing. A GPS device in a car or truck can anonymously send back streams of information about where that vehicle is and how fast it's moving. INRIX signed deals to get that input from 1.5 million trucks, delivery vans, and other fleet vehicles armed with GPS. The company also offers a mobile app for iPhones and Android-based phones. Anyone who downloads the app gets INRIX traffic information, while the phone sends INRIX its location and speed—anonymously, so no single person is tracked.

By watching what's happening all the time and processing it through a model of how traffic works in a region, INRIX can look at where any one driver is at the moment, look ahead to where she's going, and tell her how long it will take to get there.

If, say, an accident happens up ahead, INRIX will see the decrease in traffic speed and let the driver know—and help her find a better route, all on the fly.

By 2010, privately held INRIX was supplying that kind of predictive traffic information to nine of the top ten GPS navigation device makers and to most cities that use electronic highway signs to give drivers updates. The INRIX system, though, isn't yet everything it could be. Predictive technology hasn't come far enough. We need some Gretzky-brain capability flowing into the GPS units in our cars or pockets—some technology that can constantly watch events, run that information through a predictive model built on memories of past patterns, and continuously look for the best next move. We need a system that can see a traffic jam coming fifteen minutes before it happens and route us around it. But even if that's not quite happening yet, the technology is moving in that direction. And INRIX isn't the only player. Cities including Singapore and Stockholm have implemented systems that monitor traffic so tolls can be charged for driving on city streets during peak times. Those systems also collect vast amounts of traffic data and run it through predictive analytics software.

One way or another, society will eventually benefit from every car having a GPS-based device loaded with the "talent" of an experienced local cabdriver who knows local traffic patterns by heart. If all cars had predictive traffic capabilities, cities would have fewer traffic jams, saving billions of hours in wasted time each year.

WHY NOTHING SHOULD EVER BREAK

A jet airliner flying from London to New York had an almost undetectable problem. Deep in the engine, a fan blade had developed a microscopic crack. No alarm sounded to tell the pilots the plane was in trouble. Technically, the fan blade wasn't broken. It operated perfectly. But it was prebroken. If it had actually broken, it might have shut down the engine or even brought down the plane.

Jet engines are monitored constantly by temperature sensors, vibration sensors, strain gauges, and other equipment. This particular plane streamed all that sensor data through software from a Chicago company called SmartSignal. (In January 2011, SmartSignal was bought by General Electric.) Using past data, the software learns what the engine's normal operations look like, and then the system watches the streaming real-time data for variations that indicate trouble. A human looking at a report of all the data would have trouble seeing anything amiss. "It would just look like a lot of noise," according to SmartSignal board member Mike Campbell. But the software can read the data and make a prediction: that engine's fan blade is going to crack if it's not replaced.[82]

The pilots flying from London landed in New York and took the plane to maintenance. Maintenance crews inspecting the engine didn't see any problem right away but ultimately found the weak spot.

"A jet engine is the most well-understood mechanical system ever designed," says SmartSignal founder Gary Conkright. That made jet engines a perfect proving ground for two-seconds-ahead technology. SmartSignal's software can use data to inform an efficient software model of how a healthy jet engine should operate. The model can then watch the incoming data in real time, diagnose the engine while it's running, and predict failures before they happen. The technology means that no jet engine should ever break.[83]

In fact, SmartSignal is taking that idea to other realms. "Most intelligence is looking in the rearview mirror," SmartSignal CEO Jim Gagnard told us. "You are taking data and asking, 'What went wrong?' You are trying to figure out how not to do that again. There's nothing wrong with that, but we try to get the customer to look out the front windshield and see what's coming, rather than try to figure out what happened."

SmartSignal can spot problems up to six weeks before they happen. Usually these are normal examples of wear that can be solved with maintenance, allowing companies to plan for shutdowns, rather than having to take a unit out of operation without warning.

The technology has its roots in Argonne National Laboratory, which is managed in part by the University of Chicago. "Argonne ran a light breeder reactor in Idaho," Conkright explained. In the early 1990s, despite all the advanced instrumentation and smart people running the plant, he says, "during a test, they had what they categorized as a disaster. A critical feed water pump the size of a Volkswagen failed catastrophically. It could have been a disaster on the scale of Three Mile Island. The Nuclear Regulatory Commission came down hard on them."

The lab realized it had to find a way to make sure the reactor never came close to breaking again. "They found that their instrumentation had (data) archives, and they realized that if they knew what they were looking for, they could have found it in the data," Conkright said. The lab built and patented the software to do that, eventually spinning out SmartSignal as a company.

SmartSignal more recently shifted its focus from jet engines to the power and oil and gas industries. Power-generation companies generally have a fleet of plants that are differently constructed and operate under different conditions. Companies used to manage each plant individually. But in recent years energy companies have started thinking they could learn from the data about their plants to prevent them from breaking. Entergy, which provides power to 2.7 million people in Louisiana, Texas, Arkansas, and Mississippi, uses SmartSignal to oversee thirty-three different power-generation units. "In six years, they have had one thousand different incidents they caught, and all could have been problems," Gagnard said. One of those was significant: SmartSignal's software showed things were not quite normal in a 411-megawatt steam turbine-generator at Entergy's thirty-three-year-old Waterford nuclear power plant in Louisiana. The rotor of a twenty-ton generator was cracked, something no human would have detected. If it had broken, pieces of the generator would have torn through the plant at high speed. The repair cost five million dollars. If it hadn't been caught, it would have cost thirty million dollars, in addition to the potential injuries or deaths.

The next level of technology for SmartSignal will be what the company calls "prognostics"—the ability to say, with great confidence, not only what will happen and why but also when it will happen. In other words, it's very much like the concept of the two-second advantage. "We have looked at two hundred million run-time hours of equipment data to build our knowledge base," marketing vice president Joseph Dupree told us. "In order to continue to improve the prognostics, we want to still capture the data and what actually happens, to feed that knowledge and intelligence downstream. The power-generation industry has had its equipment on sensors for several decades. Now the sensors are getting better and more prolific, and that data is sitting there, ready to be mined and spotted in algorithms that can spot these failure signatures."[84]

Technology such as SmartSignal's will spread to other systems, from cars to home heating systems to the structure of a bridge. "The real barrier is that not everything that could be monitored is monitored," Campbell, the board member, told us. But that's changing quickly. And as that happens, the two-second advantage will eradicate the phrase "If it ain't broke, don't fix it." The new version will be more like "Fix it just before you know it's going to break."

As the East Orange police discovered, implementing two-second-advantage systems is one achievement. Rethinking the organization and making it into a talented enterprise takes the idea to a whole different level. Few companies have gone further down that path than Caesars Entertainment, which until November 2010 was known as Harrah's Entertainment.

Like the East Orange Police Department, what was then Harrah's had the advantage of a crisis to help convince the organization to embrace a new way of thinking. In 1998 Harrah's hired a new chief operating officer, Gary Loveman, who had been a professor at Harvard Business School and for two years worked at the Federal Reserve Bank of Boston. Harrah's had long done business pretty much the way every other gaming company did business. But other gaming companies, particularly in Las Vegas, were building opulent palaces and drawing away Harrah's customers. "We had to compete with the kind of place that God would build if he had the money," Loveman told an interviewer. Harrah's didn't have the capital to compete that way. But it did have something else of value when Loveman arrived: lots and lots of data.[85]

Harrah's offers a loyalty card, called Total Reward, that it first introduced in 1997. Customers use the card whenever they gamble, buy meals, or do almost anything at a Harrah's property, and in return they receive points that can be cashed in for rewards such as a free meal. The cards record every customer event and dump the data into a database. When Loveman came aboard, he inherited the beginnings of a vast database of customer preferences and actions. Loveman pushed his team to analyze the data to learn what makes customers happy and unhappy and what rewards they value. He wanted to know which customers were more valuable. Harrah's unearthed lots of counterintuitive insights, like the fact that it makes more money from elderly slot machine players than from any other demographic in the casino—even the high-rolling "whales" casinos traditionally spend fortunes luring. "The slot player was the forgotten customer," Loveman told *Bloomberg BusinessWeek*. "I can take you to a casino that would have a lot of young beautiful people in there and you would say, 'Man, this is a happening place.' I could take you to another place where there are a lot of people who look like your parents. The latter would be a lot more profitable than the former. My job is to make the latter."[86]

The data showed that certain customers have certain thresholds of losses, at which point those customers usually get up and leave the casino. Like Sam's Club, Harrah's found it could build models of individual cardholders, knowing perhaps that Joe Schmo likes to bring his family of four and stay at the hotel. He likes to play craps, and if he loses three hundred dollars, he leaves the casino, probably disgruntled. If he wins some money or loses less than three hundred dollars, he treats the family to dinner at the casino's buffet and ends the night taking them to a magic show at the casino—probably feeling happy. As long as Joe keeps using his Total Rewards card, the casino can track what he's doing in real time.

That kind of data let Harrah's leap into two-second-advantage technology. As Joe sits at the craps table, events move too quickly to constantly run Joe's actions through a huge database of tendencies and preferences. Instead, the system builds a fleet little model of Joe—getting to know him the way a friendly proprietor might know customers at a tiny, old-time casino. The model understands a lot about customers in general from the database, but that knowledge gets chunked into some instantly recognizable patterns, while the vast majority of data gets pushed to the background. The model learns Joe's particular tendencies and preferences and watches for triggers. So let's say Joe reaches $250 in losses, and the next roll of the dice could knock his losses up to $300—the point when Joe usually quits. The system recognizes what's going on and alerts a customer-service employee on the casino floor. The system knows that Joe often takes the family to the magic show. It also knows—from monitoring all the commercial activity in the building—that there are plenty of unsold seats for that night's show. So the system alerts the customer-service rep to tap Joe on the shoulder and offer him four free tickets to the show. Joe might never know why he got the tickets. But just as he's about to leave as a disgruntled customer, he has a reason to stay and play more—perhaps even lose more—and feel good about the Harrah's property.[87]

The system sees events in real time and can make accurate short-term predictions about what's going to happen and what action to take. Caesars Entertainment still finds a great deal of value in deeply mining databases to discover large trends, but the two-second-advantage system is all about making small, intuitive predictions just a little bit ahead of time based on events happening right then. A data-mining approach is like a hockey coach studying video of other teams to look for general tendencies to exploit. The two-second-advantage approach is like Gretzky anticipating events during a game and moving to the puck before anyone else knows what's going on.

As Loveman put this system in place, he set out to change the company so that it reacted to customer events more like a talented proprietor and less like a big gaming organization. "If you are going to do this, it has to be something that will really change behavior," Loveman said. "Some businesses are not able to cross that threshold." While other casinos spent big on opulence, Caesars Entertainment invested in information technology. Incentives were altered. The staff and executives, who had long been rewarded based on financial performance, were told they'd get bonuses based on customer satisfaction. Company policies had to change too. Caesars properties had long offered standard deals and incentives to certain classifications of customers. That was altered so that rewards could be individualized.

The technology-based strategy turned the company from an also-ran in gaming to a power player. Loveman was made CEO in 2003, and he built the company from what was Harrah's—a regional operator of fifteen casinos—to Caesars Entertainment, operating thirty-nine casinos in the United States and thirteen overseas. The company in 2011 owned the Harrah's, Caesars Palace, Bally's, and Planet Hollywood brands. In 2008 private equity firms Apollo Management and TPG Capital took Harrah's private, and the company's equity is estimated to be worth more than seven billion dollars, following a near-collapse of consumer spending after the financial crisis, versus three billion dollars in 2000 when consumers were spending like crazy. The years 2008 to 2010 were difficult for casinos and Las Vegas, but Caesars Entertainment's drive to create a predictive, talented enterprise helped it immensely.

For most of their existence, U.S. electric utilities were managed by making long-term predictions and expecting slow change. A power company generated electricity and sent it down the lines to business and residential customers, who used it in pretty predictable patterns—more during the day than at night; surges on hot summer days when air conditioners kicked in.

But in the twenty-first century, much is changing, set in motion by deregulation of the power industry. The boundaries between a utility and the electric grid have gotten blurry. More consumers and businesses are installing solar panels, which alter some of the usual dynamics, lowering demand for electricity on sunny days but increasing it on cloudy days. On long, sunny summer days, customers might even put electricity back into the grid, and they'll want to get paid for it. A sudden boom in electric cars could shift demand as people charge the cars at night.

This is what's driving so many "smart grid" projects around the United States. Utilities need to put some smarts in their infrastructure to collect and send back data so the utilities can begin to understand shifts in demand and how to meet them. That means putting sensors in the grid and smart meters in homes and businesses. A smart meter in a home can constantly measure electricity usage and display the readings and trend lines on a Web site for each homeowner. That way the homeowner can make better decisions about energy use. The information also gets sent back to the utility, giving the utilities databases full of information they never had before.

"So we have all this data—now what do we do with it?" Randy Huston, an executive at Xcel Energy, told us. "The information is interesting, but not effective until we process it and deal with it."[88] Xcel leads a consortium that's building SmartGridCity in Boulder, Colorado, to try to figure out what data can mean for a utility and its customers. Part of the challenge is to see if the grid itself can become predictive. The idea is for the system to learn on its own about demand and then watch for events—whether a downed line from a destructive snowstorm or an unusual number of air conditioners turning on during a heat wave—and instantly make adjustments. Xcel says SmartGridCity "can proactively monitor the grid's health and detect outages before they occur."[89] It's not easy getting there. One challenge is that tens of thousands of home meters are now constantly sending information to Xcel's computers, when in the past the meters would get read once a month. The data is both helpful and overwhelming. Another interesting challenge is trying to understand how quickly Xcel wants to react in certain situations. Before deploying smart technology, twenty minutes might pass between the first sign of an outage and the time Xcel could send a truck to try to fix it. Many times, the problem would clear up before the trucks rolled, Huston said. Moving trucks too soon might get costly.

The project is still in early stages, and Xcel has run into some public resistance as costs have soared. But Xcel hopes SmartGridCity will be a proving ground for a predictive future—a future where energy is saved by managing it better and no one ever experiences a power outage.

In another part of the power industry, PJM Interconnection is a regional transmission organization (RTO) that manages the largest centrally dispatched electricity grid in the world. Serving utilities that touch about fifty-one million people in the United States, PJM dispatches 164,905 megawatts of generating capacity over 56,250 miles of transmission lines as it coordinates the movement of wholesale electricity in all or parts of thirteen states. In other words, unanticipated outages could have serious if not disastrous consequences.

Until recently, when a PJM transformer failed, it would take at least five minutes before that failure was identified to the engineers and dispatchers in the operations center. In those five minutes, the surge in other transformers could cascade into blowouts that might shut down power to entire regions. The best way to avoid failures would be to notice that they were just about to happen and take action. PJM invested in a system that watches activity on the grid and constantly learns what causes problems. It builds a model that starts to get predictive—it knows that if a particular station gets overloaded and demand swings up nearby, a circuit on the other side of town is going to blow unless power gets rerouted before overall demand spikes in midafternoon. In this way, the system gets ahead of problems and fixes them before anything ever happens. It's not just connecting the dots but finding the dots in a thirteen-state grid, seeing patterns, making predictions, and taking action.[90]

At consumer products giant Procter & Gamble, the shift to a predictive, talented organization is "twenty percent a technology problem and eighty percent a cultural shift," Guy Peri, director of business intelligence at P&G, told us. Historically, the company has been backward looking. "The data comes in, gets reviewed by a marketing director, then a vice president, then on up the chain." The company in the past had been as far from real time as you could get.[91]

But tech-savvy Bob McDonald, who was made CEO in 2009,

and CIO Filippo Passerini are driving massive change. "We're looking at how to move an eighty-billion-dollar (in annual sales) elephant so it's more agile and responsive, like a five-billion-dollar gazelle," Peri said. "We're moving from the rearview mirror to forward-looking projections. We like to say we're digitizing from molecules to the shelf. P&G thinks that if we get this right, it will be a competitive advantage."

P&G markets dozens of the world's most recognizable brands, including Pampers, Tide, Mr. Clean, Pepto Bismol, Crest, and Old Spice. Through much of the 2000s, P&G had double-digit growth, and it bought Gillette in 2005 for $57 billion. Toward the end of the decade, growth slowed in the wake of a drop in consumer spending after the 2007 financial crisis. Enter McDonald, a West Point graduate who served as a U.S. Army captain before joining P&G in 1980. He became vice president for global operations in 2004 and COO in 2007. He knew the culture well enough to know it had to change for the twenty-first century. The first step was getting real-time data from retailers instead of long-after-the-fact reports. The next step shook the P&G culture: the company made the data available on a dashboard to lower-level managers and upper-level executives at the same time. The data can show performance of brands in near real time and make projections of brand market share around the world. The lower-level managers had been used to being able to view, filter, and justify such results and then pass them up the ladder. Now all the levels at P&G can see the data at the same time.

"We're driving cultural change as we're bringing on the technological capabilities," Peri said. People lower in the organization might, in such situations, be concerned about the technology, but "we want to make sure it's seen as an enabler. We're giving people information so they can make the best decisions," Peri added.

Once P&G gets the right data streaming to the right people at the right time, it can take the next step toward the two-second advantage—making P&G predictive. Software can start to learn from the data and understand when, where, and how products sell. Then P&G can do more than just feed raw data to employees—it can play off events and anticipate what's next. That might include figuring out what product to rush into certain towns after a snowstorm or understanding that a man who buys Tide and Pampers would be susceptible to an offer for a discount on Old Spice.

"We're not essentially going to run our company on autopilot," Peri told us. "The best technology plus the best minds is what we want—building systems and data can help streamline how we run the business. Our whole business model is how to take these mysteries (i.e., unsolved and unstructured business questions) into code and algorithms and free up our resources to focus on the next mystery."

A predictive, talented system becomes a part of a team effort—not a replacement for the team. Procter & Gamble sees the two-second advantage as a next-generation strategy.

Forward-thinking airlines are also moving toward event-driven, predictive systems. Today they struggle to manage mountains of data streaming in from a number of sources. Each of an airline's hundreds of aircraft sends back data about its engines, position, speed, and operations. The airline monitors weather data from around the world. It gets constant updates about ticket purchases and seat selections from its reservations systems. Each airport's ground operations sends in data about passengers, bags, gate availability, and air traffic control operations. Yet airlines can't coordinate all of that data to get a clear view of an individual customer. An airline can't, for instance, use real-time data to know that you got on a flight to Chicago but your bag didn't make it, and then automatically send you an e-mail or text message while you're still on your flight telling you where your bag is and when it will be delivered to you. And that's on the simple end of the predictive systems spectrum.

Air France and Southwest Airlines are among the airline companies driving toward two-second-advantage technology. Air France is putting in place a system that can track passenger flow among all its aircraft and connection points, said Air France's chief information officer, Edouard Odier, when we talked with him. On long-haul flights, as many as 40 percent of the passengers will be coming into the airport on other flights and connecting to get on the long route. Air France's system will start off by building agile models of the passengers scheduled on those long-haul flights and watching what's happening on those passengers' connecting flights. "If there are four hundred people on the long flight and the system sees that three will miss it, it may decide it's better to leave and have four hundred on time and three left behind," Odier said. "But if forty are on connections running late and will miss the flight, it might be better to wait for them." The idea is that if the system helps to decide to let the flight take off, it will know who will miss the connection and automatically e-mail them with rebooking information.[92]

"The key for us is less data, not more," said Jan Marshall, CIO of Southwest Airlines. "We want a system that's almost innate." Her team has been working on developing the rules and logic that have become intuition to Southwest's dispatchers, crews, and other operations staff. The plan is to build those rules into technology that can watch all of Southwest's data streams and chunk airline operations well enough to help the airline get ahead of events like storms or aircraft maintenance issues. The domino effect of such events is hugely complex, yet there's no time to analyze mountains of data to arrive at an optimal plan. The system needs to be able to stream data about events through a model that can instinctively and quickly offer up the best options. Down the road, Marshall even envisions a smart-phone app that becomes part of this system. The app would know a passenger's flight information, and the GPS-enabled phone would know where the passenger is and (if the passenger opts in) let Southwest know too. Southwest could know, for instance, that a passenger scheduled for a flight due to leave in twenty minutes is already in the airport and perhaps snaking through security—and might choose to leave the aircraft's door open a little longer so that passenger can get on board.[93]

The United States is a nation of immigrants. People come into the country from all over the world, with all kinds of different circumstances, but just about every immigrant to the United States in the past thirty years has shared a common experience: a frustration with the long delays, mixed-up paperwork, and overextended agents that make up the federal immigration bureaucracy. A year after the terrorist attacks of September 11, 2001, President George W. Bush created the Department of Homeland Security and pulled together all of the responsibility and functions for processing immigrants into a new bureau called United States Citizenship and Immigration Services (USCIS). As a birthday president, the USCIS got a backlog of 3.5 million applications for citizenship or residency. The information about those applicants was stored in twenty-three legacy computer systems that, for the most part, couldn't communicate with one another.

Meanwhile, seven million new applications poured in each year. All the while, the bureau was under pressure to make sure it kept potential terrorists out. Far from having a two-second advantage, the agency was more like two years behind.

In its early years, the USCIS tried to fix its tangled process and mostly failed. A Homeland Security report in September 2005 has a telling title: "USCIS Faces Challenges in Modernizing Information Technology." The report details a technological mess and an inability to think about the agency's information systems in a new way. "USCIS uses multiple, disparate information systems that are difficult to use and do not adequately share information, resulting in data integrity problems," the report stated. "Despite federal requirements, USCIS has not had a focused approach to improving the processes and systems used to accomplish its citizenship and immigration mission."[94]

A few years later, a new idea started to percolate through USCIS. It took hold in pockets of the bureau and began to expand. The idea was this: the way to fix the technology morass is not a complete overhaul of the technology; it is an overhaul of the way USCIS thinks. Before the late 2000s, federal immigration agencies had a form-centric approach. In other words, everything revolved around certain forms that immigrants had to fill out. One kind of form was stored in one kind of database with one kind of structure, while another form was in another database with another structure. (Some forms never made it into electronic data—they just sat in boxes!) The forms weren't linked or cross-referenced. If an agent wanted to check the status of an immigration applicant, she'd have to log in to different systems to search for different forms—or in some cases literally slide her chair over to another computer terminal. Obviously, the form-centric approach no longer worked.

So USCIS has slowly pushed its culture to a person-centric approach. "We're in the middle of a huge transformation," said Leslie Hope, deputy chief information officer at USCIS. "It will change everything." The person-centric approach organizes the forms around a person, instead of the other way around. There's even a name for it at USCIS: Person Centric Query, or PCQ. The idea, obvious as it might seem, is to create a system that allows an agent to call up an applicant's name and see all that person's forms and information from all the different systems gathered into a single, easy-to-navigate view on-screen. "Instead of chasing data files and paper all over the country, the new system not only will allow us to centralize all that business around a person, but it will include biometrics and any other data related to that person from start to finish," Hope told us.[95]

How is USCIS doing that? It could reconfigure and migrate all the data on all the different legacy systems into a single new system, but that approach would take years and probably would result in lost data and errors. Instead, USCIS has added a layer of software—a software bus—between the human agents and the many legacy computer systems. The bus acts as a translator and data-collection helper. When an immigration agent calls up an applicant's file, the Person Centric Query software pings twenty-three different USCIS systems, grabs the data about that applicant, and reformats the data so it can all be presented together. The agent can for the first time see a complete administrative picture of an applicant, updated to the moment the agent is looking at the screen. USCIS employees can then start thinking of their jobs as serving people instead of serving forms and a bureaucracy—and that seems to make a world of difference. USCIS leadership phrased the new approach this way: "USCIS will deliver the right benefit to the right person in the right amount of time, while ensuring that the wrong individual does not access immigration benefits."[96]

The shift to a person-centric approach is a work in progress and probably won't be completed until 2014, Hope told us. But USCIS is betting that it will result in a better experience for immigrants and better security against people the United States doesn't want inside its borders.

As it stands in 2011, the person-centric technology is set up to respond to queries from USCIS employees—which is still more of a twentieth-century database-style approach, not a two-second-advantage approach. But this is where the USCIS's shift gets interesting. If the organization can embrace the person-centric culture, the technology would be in place to help USCIS get more proactive and predictive. The next step would be to make the person-centric technology smart enough to build the equivalent of mental models of each immigration applicant. Instead of only responding to an agent's query, the USCIS system could turn into an advocate for the applicant. It would know what forms and steps the applicant needed to address, for instance, and send alerts to the applicant. On the other end, when an applicant had completed necessary procedures, the system could proactively let agents know. There are all sorts of events that might affect an individual applicant, including a change in job, a marriage, an arrest, or a change in immigration laws. The person-centric system would instantly know what each event would mean to the applicant, much as if an experienced agent were watching over that particular applicant's case. The system would proactively know what the applicant needed to do next based on the event that just happened. At the same time, if the system recognized events that indicated trouble, it could proactively identify immigrants who could be dangerous.

If the person-centric system evolves this way, it will ease the demands on USCIS agents, cut down the backlog of cases, and vastly improve the immigration experience for people who move to the United States. It's happening slowly, as things do in the federal bureaucracy. But as Hope told us, the future of USCIS is riding on becoming person-centric—and, eventually, predictive. It's certainly an example of how both a culture and the technology have to change to get the two-second advantage—and how that change is sometimes hard.

In the first part of the book, we talked about how talented people have an ability to predict what's going to happen a little faster and a little better than almost anyone else. They have what we're metaphorically calling the two-second advantage. Some of these people, like Eduard Schmieder as a young violin prodigy, seem to get that capability largely from their hardware—in other words, their brain's wiring. Others, like Boston mayor Tom Menino, rely on remarkable software—or mental models—that they've built into otherwise typical brains. Hockey great Wayne Gretzky seems to have been blessed with both.

We then discussed how technology has to change and work more like talent in brains. The long-standing technology model of managing information with database and analytic software increasingly can't keep up with the onslaught of data and fast pace of events. Next-generation technology has to use efficient models that can take in a stream of events, constantly make instantaneous and highly accurate predictions about what's likely to happen, and act on those predictions. Technology has to learn from the mass of data and then set it aside or even forget it, much as brains do.

Some pioneers are implementing this model of technology,

including companies like Sam's Club and INRIX, and the new technology can create a new cultural mind-set inside organizations. Whole organizations can work toward reacting to events more like a talented human, instead of like a lumbering bureaucracy. Procter & Gamble is investing in that direction, as are entities as disparate as Caesars Entertainment and the East Orange Police Department. These entities believe they can develop an organizational two-second advantage.

The technology behind the two-second advantage is in its earliest stages. Technology available in 2011 can track events from dozens of sources in real time, as they happen. And it can run those events through narrowly focused rules-based software that can instantly make a calculated prediction—the way Reliance in India can anticipate when a customer is about to get angry and drop it for another carrier.

Events, which come in from all types of sources (a Web site, a sensor in the field, an office PC, a camera) can be collected through a software bus, the way the Bureau of U.S. Citizenship and Immigration Services is doing it. The bus can translate the data from the events into a format that can be processed by a software engine programmed using rules learned from mining the database. The idea is to avoid going back to search the old data in the database and instead process the events through the rules, which can come to instantaneous conclusions. The rules may be set up to take into account individual situations, the way the Harrah's system can "learn" about a customer's preferences and tendencies by checking past data in the database.

To some extent, technology can tune the rules itself by watching outcomes and altering the rules based on those outcomes. There is a term for this: "forward-chaining." The system starts out with rules programmed into it and creates concepts from those rules. If the rules expect *X* to happen but instead *Y* happens, the rules treat *Y* as the new expected outcome and move on to build a more refined concept. This is a beginning step toward brainlike processing and can be put into practice today.

Algorithms—the calculations behind the software—keep getting better at making real-time predictions. Cloud computing—the ability to pool computing resources through the networks—makes it easier to collect events from disparate sources and add memory for running the rules-based software when needed. Visualization technology is giving humans better ways to work with real-time, predictive computing to make better and faster decisions. The East Orange police built their own dashboard. Procter & Gamble and others are buying theirs from major technology providers.

Today's technology can react and predict in ways that are becoming intriguingly brainlike—as long as humans set it up specifically to do those things. People have to tell the machines what events to watch and what rules to follow. The machines can't yet really learn. They can't decide what to remember and what to forget. The machines still operate like computers, processing information the same way they have since the 1940s—though of course much, much faster.

That's the great barrier to getting to true, two-second-advantage, talented systems. Faster is no longer an adequate solution. Computers and software will have to work in an entirely new way—a more brainlike way. That's the next frontier.

Brainy Electronics and Electronic Brains

In the summer of 1955, John von Neumann started feeling shooting pains in his left shoulder. Living in Washington, D.C., and working as a recently appointed member of President Eisenhower's Atomic Energy Commission, von Neumann was just about to begin a prestigious lecture series for Yale University called the Silliman Lectures. For more than two decades, he'd been one of the most influential mathematicians in the world, contributing to game and economic theory, quantum mechanics, and other scientific fields. In the 1940s, von Neumann had developed an expertise in the mathematical modeling of explosions, which had led to him being recruited for the Manhattan Project, where he had helped research and construct the first atomic bomb.[97]

Von Neumann had always worked zealously, diving headlong into new research areas in which he had no previous training.

In the 1940s, hearing about budding work on the first electronic computers, he had become convinced that their superfast calculations could help solve difficult scientific problems—including the modeling of explosions. So he went to visit J. Presper Eckert and John Mauchly, who were building the Electronic Numerical Integrator and Computer (ENIAC) at the University of Pennsylvania. Before long, von Neumann was helping Eckert and Mauchly develop ENIAC, and in the process von Neumann described a computer architecture that was virtually unheard-of at the time. Computers to that point had been programmed by physically altering their circuitry. Von Neumann proposed storing the program in memory alongside the data. The program would fetch the data, run the information through the program's algorithms, spit out a result, and fetch more data to continue the process.

Of all of the things von Neumann accomplished in his life, this is the one he's best remembered for. The architecture he outlined—which came to be known as the von Neumann architecture—is the way nearly every computer ever built has worked, right up to modern times.

As the pain in von Neumann's shoulder worsened, he had surgery and was diagnosed with bone cancer. He knew he had limited time left, so he dove into writing the manuscript for the Silliman Lectures. He wanted to write about a topic that had been on his mind for some time. The title of the manuscript was *The Computer and the Brain*. To give you some idea of von Neumann's intellectual courage, the first words of the manuscript are "Since I am neither a neurologist nor a psychiatrist, but a mathematician, the work that follows requires some explanation and justification."

Von Neumann wrote about one hundred pages of the lecture. The text started out describing how computers work and then moved into what was known in the 1950s about the brain's mechanisms. Astoundingly, von Neumann made a point that echoes more strongly in the 2010s than it must have in the 1950s. He compared the kind of fuzzy, parallel, simultaneous processing that goes on in the brain to the serial, instruction-based processing of computers. "It is to be expected that an efficiently organized large natural automaton (like the human nervous system) will tend to pick up as many logical (or informational) items as possible simultaneously," von Neumann wrote. "While an efficiently organized large artificial automaton (like a large modern computing machine) will be more likely to do things successively—one thing at a time."[98] In the brain, instructions and data look like the same thing, and the connections between them are many and ever changing. In computers, instructions and data are separate, brought together through the single pathway of the microprocessor. As much as anything else, von Neumann concluded, this is the fundamental difference between how brains and computers function—and ultimately the barrier to making computers more like the brain.[99]

Von Neumann never finished writing the lecture and never delivered it. In April 1956, he was admitted to Walter Reed hospital in Washington, his cancer worsening. He tried to work on the lecture while there but could not manage it. He lingered until his death on February 8, 1957.

The irony is that von Neumann seemed interested in moving computers beyond his own architecture to the next level—studying the brain for ways to make better computers. Fifty-five years later, the problem von Neumann began thinking about has not been solved. The von Neumann architecture has come to a moment when it's referred to as the von Neumann bottleneck. Computers cannot work much faster using the same serial architecture, yet the flood of data and quicker pace of business demand faster processing. Now, in the 2010s, there's an explosive new energy behind the work of solving von Neumann's problem. A new, post–von Neumann architecture will be necessary to developing a next generation of computers that can keep up with the world's complexity. It will also be important to the two-second advantage and talented systems.

This works the other way around too. Research into how to make computers more brainlike has become critical to understanding the brain. "The computational theory of mind has quietly entrenched itself in neuroscience," wrote Steven Pinker, a Harvard University psychology professor. "No corner of the field is untouched by the idea that information processing is the fundamental activity of the brain."[100] In such a milieu, scientists find it helpful to discover the differences in how information is processed in computers versus brains. The findings should help us build both brainy computers and better minds.

In the middle of the classic 1984 movie *The Terminator*, Sarah Connor is searching through a phone book with Kyle Reese. Kyle has been sent back from the future to protect Sarah, who will become the mother of John Connor, the man who, in the next generation, will lead the human race in its fight to keep from being annihilated by a race of intelligent robots. Kyle and Sarah are being chased by Arnold Schwarzenegger as the Terminator—a brutal assassin cyborg sent back from the future to kill Sarah and prevent John from being born. Sarah is looking through the phone book for the address of Cyber Dynamics Corporation, the company that will build the computers that will evolve the first artificial intelligence that can learn and then replicate. The company's technology will initiate the demise of humans. Sarah finds the address.

"Isn't that it? Cyber Dynamics Corporation?" she says to Kyle, in the script by James Cameron.

"What about it?" he asks.

"Didn't you say they're going to develop this revolutionary new thing?"

"Molecular memory."

"Whatever. They become the hotshot computer guys so they get the job to build El Computer Grande—Skynet—for the government, right?"

"That's the way it was told to me," Kyle says.[101]

She proposes attacking the company, but of course Sarah and Kyle never manage to stop Cyber Dynamics from building an intelligent system, and the humans' dismal future inevitably unfolds in the story.

The movie is based on a familiar science-fiction trope. Since at least the 1950s, when the popular press referred to computers (which at the time had less processing power than a present-day digital watch) as "giant brains," people have feared that machines would soon outsmart us. And now, in the 2010s, neuroscientists and computer scientists are enthusiastically working to create machines that operate more like human brains. The scientists even have a goal of building systems that have *talent.* If humans haven't yet created technology that threatens to make *The Terminator* come true, surely now we must be playing with fire, right?

Well, not exactly. Scientists are working mightily to borrow ideas from the brain to give computers human, brainlike qualities. But something interesting is happening in that quest. As neuroscientists learn more about the brain, they're finding out that it is even more complex and difficult to duplicate than ever imagined. And as computer scientists get closer to replicating brain qualities in machines, they're finding out just how daunting that task is. Just think back to scientist Henry Markram's rant about cat brains in chapter 4.

It's probably helpful, then, to discuss a few of the enormous challenges to getting machines to work like fully functioning brains—and some of the successes in doing the same. That should help lend some perspective on the advances scientists and companies are making in this field. Building the first talented, two-second-advantage systems is not the same as the stupendously difficult task of building thinking, sentient machines.

Scientists have long been overwhelmed by the sheer density of the circuitry in the human brain. A lump of tissue you could hold in one hand contains around one trillion neurons connected by ten quadrillion synapses.[102] The collection of neurons in a single brain can hold about a petabyte of information[103]—equal to about twenty million four-drawer filing cabinets filled with pages of text.[104]

Oddly enough, the brain's basic hardware is one of the lesser hurdles for technologists. If we make the assumption that a neuron is roughly the equivalent of a single transistor, then a trillion transistors is not that spectacular. You could find more than a trillion transistors in the computers in a typical college dorm building. The density of transistors packed on to computer chips today is about ten times less than the density of neurons in the brain, and while engineers in the 2010s are having a harder time improving chip density at the same pace as in the past, there's no doubt density will continue to get closer to the neuron density of the brain. So at face value, the difference in raw processing capability between computers and brains will eventually be pretty small.

As for storage, a petabyte is a lot. Google was processing about twenty petabytes of data a day through its global network in 2010, so imagine a brain holding one-twentieth of Google's daily data stream—maps, searches, photos, videos, documents, etc. Yet a one-petabyte hard drive will probably be available on the market by the late 2010s. New technologies such as IBM's experimental "racetrack memory" are expected in the coming decade to surpass the storage density of the brain.[105] In the 2020s, people could be keeping petabytes of data on a gadget the size of a USB drive. So the gap between computer storage and brain storage is closing too.

However, simply matching the brain circuit for circuit won't create a brain. One of the remarkable features of the brain is how it processes so much data through all those neurons on so little energy—about twelve to twenty watts, or the amount of power it would take to run a refrigerator lightbulb. If a computer built with 2011 technology processed the same amount of information at the same speed as a single brain, the computer system would consume as much energy as a small city. Computer technology is an energy hog compared to brain technology. Unless that changes, the Terminator would have to drag around a nuclear power plant behind it just to think.

The human brain's quadrillion synapses link every bit of data in multiple ways. (Consider all the associations that pop up just from seeing the word "chocolate.") No programmer could be smart enough to think through and create all those links. The synapses wire themselves together based on experiences, so each brain programs itself. That's why we're all unique. No one knows how to make a computer do this.[106]

Memory and processing are essentially the same thing in brains, performed in the same place. This is a reason the brain is so energy efficient. Brain architecture is fundamentally different from von Neumann's architecture, where memory is in one place, the processing is elsewhere, and data gets sipped by the processor through a relative straw. The cerebral cortex—that crucial outer layer in the brains of mammals, where most learning and processing takes place—looks the same everywhere: a layer of neurons wired to one another by synapses. The brain learns cause-and-effect relationships and forms associative memories using a single elegant process.

Recall from earlier in the book that the lowest layer of neurons receives signals from the outside world. If you imagine a baby looking around at the world, her brain is more or less a blank slate. Neurons just sit there, waiting to receive input. When she looks at a picture of a frog, her eyes send a stream of data to the part of her cortex that processes visual signals, and the neurons in that region take note of the characteristics of a frog, broken down into discrete pieces of data. If the baby's parents say the word "frog," neurons connect that sound to each characteristic of a frog. In time, the brain creates a folder called "frog" and puts all information about frogs in it, creating a set of associative memories. The brain is basically wiring itself—making connections between neurons based on experience. Those synaptic connections get reinforced through the repeated firing of neurons, developing into chunks. Soon any mention or sight of a frog fires up all of those frog-connected neurons in the same instant. It's a massively parallel process. "Frog" lights up all frog data at once and then focuses on what's most appropriate for the moment.

To go back to the description by neuroscientist Jim Olds,

your brain thinks a lot of thoughts at the same time and then gravitates toward a solution or idea. The brain's programming is fuzzy and flexible. Computers, by contrast, process everything serially, calculating their way to a conclusion. Computer architecture dictates a tedious, rigorous process, causing the von Neumann bottleneck. Though much work is being done to come up with a post–von Neumann architecture, nothing has yet proven successful.

Emotion creates another hurdle for brainy computers. Science fiction often conjures up intelligent robots that have no emotion. The reason given in these stories is that emotion is too hard to program into machines. But there's a problem with that logic. Emotion is an important part of the brain's efficient processing and memory capability. Emotion is an essential information cross-indexer. It colors every bit of information and helps tell your brain how important certain information is or how it relates to other information. It would be extremely hard to create a brainlike computer *without* emotion—yet no one has the slightest clue how to program emotion. This catch-22 might be enough to keep sentient machines from emerging for a long, long time.[107]

One more barrier to building a computer brain is hardiness. If a bunch of transistors on a computer burn out, the computer goes down. If your brain worked that way, a single drinking binge would give your brain the blue screen of death. Your brain has dynamic wiring—it can change the way its neurons are wired together on the fly. When a neuron dies, your brain just routes around it. If someone has a stroke or an accident and an entire region of the brain goes on the blink, the person can often regain some or all of that capability. The brain slowly but surely rewires itself around the problem. Technologists can't make computers do anything like this, and as computers get increasingly complex—with more transistors and more connections and more lines of code—they tend to get more brittle. In all that complexity, a little glitch can bring the operation to a halt.

The message here is that computers are not likely to become brains—at least not in any foreseeable future. But that doesn't mean computers can't become more brainlike. Computer scientists and neuroscientists are hacking the brain a little at a time, at the edges, approaching it from a number of different angles. "The disparity between modern computer architectures and the brain is so large that you have a large potential gain" by doing this research, says Dharmendra Modha, who is driving the cognitive computing team at IBM's Almaden research lab outside San Jose, California. "If you get even halfway, you change the world."[108] Getting computers to operate just a little more like brains would be a fantastic advance.

And that's what's starting to happen. One thrust of research is to create hardware and wiring that borrows from brains, starting down a path to give machines some of the raw talent found in people like musician Eduard Schmieder or comedian Mo Rocca.

Another thrust involves work on software, finding ways for machines to learn and build models that can quickly and efficiently react to events—the kind of capability built up in the brains of people like Boston mayor Thomas Menino or pickup artist Mystery.

Scientists and engineers are pursuing this work in labs and R&D centers. Meanwhile, companies and IT departments are implementing early versions of this technology in the field. As enterprises such as Sam's Club, Xcel Energy, and the East Orange police build and deploy predictive systems, we learn a little more about how machines can respond like talented humans. The functioning systems at this stage are narrowly focused on specific applications, like making instant offers to Sam's Club customers or anticipating certain crimes. But these are starting points on the way to building talented systems and talented enterprises. More research and development will be needed to move the world away from von Neumann's architecture and into architecture that can really drive two-second-advantage systems.

In the mid-2000s, the Pentagon's Defense Advanced Research Projects Agency, or DARPA, asked for bids to design a new type of electronics system, one that would operate less in the realm of computer science and more in the realm of brains. The program is called SyNAPSE, an acronym for "Systems of Neuromorphic Adaptable Scalable Electronics." The ultimate goal of the program is a computer chip that can act like a brain with roughly one hundred million neurons.[109] That would create a brainlike chip that falls somewhere between the cerebral cortex of a rat (fifty-five million neurons or so) and a cat (about 760 million neurons). Humans have more than twenty billion neurons. DARPA also wants to see those chips installed in a robotic platform that can perceive the environment around it, move in that environment, and learn from its experience. Kind of a tall order, if you think of the relatively small first steps involved in Rajesh Rao's robotics work.

In the past decade, DARPA commissioned two similar projects aimed at creating computers that could process information more like the brain does. One was called BICA, for Biologically Inspired Computer Architecture; the other was HYCS, for

High-Yield Cognitive Systems. Neither produced a post–von Neumann architected chip. But since those projects, there have been big advances in the sciences that can drive this technology.

By late 2010, DARPA had awarded SyNAPSE grants to three teams: IBM Research, HP Labs, and HRL, the former Hughes Research Labs.[110] All three must hit milestones that show their progress. They have to start by developing a simulation that mimics the neural behavior of primitive animals. The second phase calls for a computer model that can operate at the level of a mouse. Phase three calls for something at a cat level. People involved in the project stress that the goals aren't meant to be taken literally—no one is setting out to build a replicant cat. The brains are simulations, working entirely inside software.[111] Promising approaches from all three teams could receive contracts from the Pentagon to continue the research and turn it into applications.

The military sees a lot of possible uses. Some, as you might imagine, are about smarter weapons. The military's drone aircraft—one of the most significant war-fighting advances of the past decade—today are operated remotely by people on the ground. Those people, surprisingly, are almost as likely to develop posttraumatic stress disorder as pilots who fly actual missions. A fully autonomous drone would eliminate that risk and take away other limitations of human operators—like the fact that humans get tired or make mistakes.

Other applications are not so battle oriented. Brainlike technology could analyze satellite imagery. Satellites send back enormous amounts of data, and today almost all of it must be viewed and analyzed by people. Computers can handle far greater workloads than a typical intelligence analyst, evaluating more images with greater precision without tiring.[112] Computers today do some visual analysis, but they can't really do the kind of humanlike thinking that involves recognizing patterns and drawing conclusions.

Still a third use is in the realm of space exploration. Vehicles in space must receive all their operating commands from a person sitting at a computer terminal on earth. But during a Mars mission, radio communication signals traveling at the speed of light take ten to twenty minutes to go one way. If a Mars rover were headed for a dangerous crater, the operator wouldn't see the image from the rover until more than ten minutes after the rover sent it. And then, if the operator put the brakes on, the signal wouldn't get to the rover for more than ten minutes. By then the vehicle could be flipped over in the crater. An intelligent vehicle would be able to evaluate its surroundings and move about on its own. It would still have contact with people here on earth but would need less direct supervision.

IBM's work in cognitive computing started in 2006, about two years before DARPA announced SyNAPSE.[113] Since then, the team has run "cortical simulations," or computer models of the brains of increasingly complex animals. In 2006 the team used a Blue Gene supercomputer to simulate the cortex of an animal with eight million neurons and fifty billion synapses, or about half the size of a mouse brain. A year later, they designed a simulation of the rat cortex, and in late 2009 they worked their way up the food chain to simulate a cat cortex. That's the work that Henry Markram criticized.

At the Brain Mind Institute in Lausanne, Switzerland, scientist Markram is leading another effort to model brains. It's not part of SyNAPSE but instead gets backing from the Swiss government, the European Union, and IBM. (IBM, as you can see, is very into brainy computers.) Markram has in turn been criticized by other scientists for saying he can build an artificial human mind by 2018—one that might, on its own, develop some version of human consciousness. The sheer scale of such a feat seems implausible. The computers involved in brain simulations are massive—IBM's cat simulation used the Dawn Blue Gene/P supercomputer at Lawrence Berkeley National Lab, which has 147,456 central processing units (CPUs) and 144 terabytes of main memory and requires a megawatt of electricity to keep running (about enough to power one thousand homes). Even with that kind of power, the simulations processed information at a fraction of the speed of an actual brain. To create a single operational human brain on a computer would just about drain the world's supercomputing resources and bring down the electrical grid. Brain simulations are still harnessed to—and hamstrung by—traditional semiconductor design. So some research teams are trying to take what we're learning about the brain's software to build a new kind of hardware.

Back in 1971, Leon Chua, a professor at the University of California at Berkeley, published a theory that a chip could be made that would change resistance when a voltage is applied across it. He called these chips "memristors." They weren't possible to fabricate, even in labs, until scientists made discoveries in nanotechnology in the 1990s and 2000s that allowed them to manipulate atoms to make new materials. In recent years, research labs have experimented with pasting a tiny layer of titanium dioxide on a processing chip. By running current through the material on top, they can change the resistance of the elements on the chip underneath, and that change stays intact even after taking away the current. In other words, exposure to a signal changes the way the chip is wired, and the chip both processes the information and remembers it—a little like neurons in the brain. Researchers have even built a basic network of a few memristors and shown how such a device could create very primitive associative memories.

A major difference between computers and brains is that computers process information in one place and store the data in another. In brains, the cortex operates as both memory and storage in the same place, processing lots of data at the same time in parallel. Memristors might be a path to that kind of processing. And because memristors can both process and store information, they should be able to work much more efficiently, consume less power, work faster, and give off less heat than current chip and storage architecture.

The Hewlett-Packard team that's getting funding from SyNAPSE is focused on hardware—and specifically memristors. HP got interested when one of its HP Labs scientists, Greg Snider, realized that a nanoscale device he was working on exhibited behavior that had been predicted by Chua. The discovery is crucial to eventually making memristors with enough efficiency and density to rival biological computation.[114]

HP, though, is looking at taking a practical approach to reaching DARPA's goals, with plans to also develop memristors into more near-term products. "The DARPA guidelines are very broad," said Massimiliano Versace, who is a senior research scientist at Boston University and one of the principal investigators on the HP team. "What we're doing is taking those and creating more precise and limited functions."[115] In 2010 HP signed a deal with Hynix Semiconductor to develop commercial memristor products under the name ReRAM (Resistive Random Access

Memory). The first commercially available versions are scheduled to hit the market in the mid-2010s, most likely as a replacement for flash drives. But the potential is there to string together bundles of memristors and get them to behave like neural networks. Memristor-based hardware could work more like the cerebral cortex. If the memristor-based neural networks can program themselves by forming connections among pieces of data—the way brains do it—we'll see the first glimmers of hardware that can learn. And hardware that can learn would significantly help in the building of two-second-advantage technology and talented systems.[116]

HP's research revived Chua's four-decades-old interest in memristors. "Professor Chua himself pointed out the connection between the properties of his proposed memristor and those of a synapse in his earliest papers," said Stan Williams, HP senior fellow. Chua is working with HP on making memristors do brainlike processing.[117]

At Stanford University, bioengineer Kwabena Boahen is developing another step on the path to brain-inspired hardware: neuromorphic chips. "Neuro" comes from the same prefix in neuroscience; "morphic" refers to the chips' ability to morph and create new connections and pathways based on input. In other words, the chips would have the ability to learn. A key insight from Boahen is that talented brains literally change their hardware to do a specific task well. After all those years of hockey experiences, Wayne Gretzky's brain had literally rewired itself to think about hockey in a supremely efficient way. His brain's wiring would look nothing like the wiring in Mayor Menino's brain, which was wired for the task of being a Boston politician. Computers, on the other hand, are general-purpose machines.

Whether a laptop or a mainframe, the computer hardware doesn't get altered to focus on a specific task. Instead, we make the generalized computer do specific things by running software. The generalized construction of computers has a high cost in efficiency and speed. Lots of computing resources are wasted doing things that don't really matter for the task at hand.

"Customizing the hardware is something brains and neuromorphic chips have in common—they are both programmed at the level of individual connections," Boahen wrote in *Scientific American*. "They adapt the tool to the specific job. If we could translate that mechanism in to silicon—metamorphing—we could have our neuromorphic chips modify themselves in the same fashion."[118]

In 2009 Boahen's team had built a batch of its first chips, which they dubbed Neurogrid chips. Each had enough self-wiring memristor-like devices to emulate 65,536 neurons. The team has since mounted sixteen Neurogrid chips on a single board to emulate one million neurons, connected by a tangle of six billion synapses. In 2011 they hope to create a second-generation Neurogrid array of sixty-four million silicon neurons, about equal to a mouse brain.

Here's the big difference between Boahen's path to a simulated mouse brain and the path taken by IBM with its mouse brain: IBM used sixteen supercomputer racks to simulate fifty-five million neurons connected by 442 billion synapses. Those computers consumed 320,000 watts of electricity, enough to power 260 American households. (The cat brain simulation required three times that much power.) When Boahen runs a simulation of a mouse brain with a similar number of neurons, his Neurogrid device will get by on just a few watts of power.[119]

Overall, each of these advances—the simulations in supercomputers, memristors, neuromorphic chips—will help in the quest for brainlike computers.

One other interesting hardware advance is on the horizon: the possibility that post–von Neumann architecture won't involve anything that remotely resembles computer chips but instead will rely on the spin of atoms. This is the odd and elusive field of quantum computing, which delves into areas that are nearly unthinkable. For instance, it's possible that a quantum computer holds an infinite number of right answers for an infinite number of parallel universes. It just happens to give you the right answer for the universe you happen to be in at the time. "It takes a great deal of courage to accept these things," Charles Bennett of IBM, one of the best-known scientists in the field, told us when he was working on quantum computers more than a decade ago.[120]

Bennett was getting at the key property of quantum computing: it figures all possible answers instantly and then chooses the best one for the moment at hand—which sounds like the way brains work. IBM, Yale University, the University of California at Santa Barbara, and others are working on quantum computing chips that could be mass-produced using known techniques.[121] D-Wave, a start-up based in British Columbia, Canada, got funding from Silicon Valley venture capitalists back in 2003 and now is testing quantum computing–based chips.[122]

Quantum computing moved forward when physicists realized that atoms are naturally tiny little calculators. "Nature knows how to compute," said MIT's Neil Gershenfeld, who helped build one of the earliest successful quantum computers. "We just didn't know how to ask the right questions." Atoms have a natural spin or orientation, the way a needle on a compass has an orientation. The spin can be either up or down, on or off, a one or a zero (the way both computer chips and neurons represent information). But since an atom can be both up and down at once—called putting it into a superposition—it's not just equal to one bit, as in a traditional computer. It's something else. Scientists call it a qubit. If you put a bunch of qubits together, they don't do calculations linearly, like today's computers. They are, in a sense, doing all possible calculations at the same time, straddling all the possible answers. The act of measuring the qubits stops the calculating process and forces them to settle on an answer. One of D-Wave's chips, with 128 qubits, could have the power of a supercomputer.

To program a quantum computer, you wouldn't use the step-by-step logic of today's computers. You'd want logic that used the properties of qubits. That's what Lov Grover, then at AT&T's Bell Labs, did in the 1990s when he invented an algorithm that uses quantum computing to search databases. Grover's algorithm sets up multiple paths of computations, so that waves of results—all happening at the same time—interfere with one another. The right answers interfere constructively and add up. It's not all that different from the way Jim Olds described the way brains gravitate to an idea or answer.

The most intractable problem of quantum computing is that the inner workings of the device, the actual calculating of atoms, have to be completely isolated from their surroundings. Any interaction with even a single other atom or particle of light makes the particles choose a spin direction, polluting the results. And yet if you're going to program a quantum computer, put in data, and get out a result, you have to interact with the atoms somehow. Some scientists have used quantum entanglement—in which particles are linked so that measuring a property of one instantly reveals information about the other—to extract information. But creating and maintaining qubits in entangled states has been a serious challenge. Other groups have tried measuring the magnetic warble of the atoms from afar with a nuclear magnetic resonance machine.

The bottom line is that quantum computing is still a lab experiment, but across the board scientists believe it will come into use in the next twenty years, perhaps altering the field of computing as much as the invention of the transistor did in the 1940s. Clearly, at some point, computer hardware will transition into a post–von Neumann architecture.

Between now and then, though, the best hope for brainlike machines lies in software.

Jeff Hawkins stretched out in an office chair that could barely hold his lanky frame. He was in the headquarters of Numenta, in the attic of a small building in downtown Redwood City, California. A wood-beam ceiling slanted down to tiny hobbit-size windows. The lone, beat-up desk could've been lifted from a DMV. The room had harsh fluorescent lighting and no decor whatsoever. The office belonged to Donna Dubinsky, Hawkins's longtime business partner, who helped him start and run Palm Computing and then Handspring, which made the original Treo. "I don't like offices," Hawkins said. "I just find somewhere to work." Numenta seemed to be anything but a cushy glamour project for Hawkins.[123]

His 2004 book, *On Intelligence*, brought together emerging ideas and put forth a theory about how the neocortex works, and especially how it works as a predictive machine. As you'll recall, it influenced researchers such as Rajesh Rao. You'd think Hawkins would consider the book a nice achievement on the side, since over about twenty years he created some of the most impactful handheld technology in history. But actually, Hawkins believes his work in technology has always been a way to finance his brain research. "I have been a part-time entrepreneur," he said, which seems a bit like John McEnroe saying his tennis career was a way to finance his ambition to be a talk-show host.[124]

In 2002, Hawkins formed the nonprofit Redwood Neuroscience Institute (which has since been renamed the Redwood Center for Theoretical Neuroscience) to further develop his theories. There, mathematician Dileep George came up with algorithms that would allow Hawkins's theories to be turned into computer software. That, in turn, became the original core technology behind Numenta. Hawkins asked Dubinsky to be CEO. She was just finishing a year off from work and was bored. "I thought, God, I would love to work with Jeff again on something that could change the world," Dubinsky said.[125] So Hawkins, Dubinsky, and George cofounded Numenta. The company's goal from the start was to make computer technology that exhibits brainlike capabilities.

"I want to be a catalyst for a future industry of machine learning based on biological principles," Hawkins told us. "I believe there's a monstrous opportunity for machines that can learn like brains. It could change the world in a big way. But it's a great intellectual challenge."[126]

In fact, it's such a tough problem that when we talked to Hawkins in 2010, he said of Numenta: "It's been five years, and we have not been successful."[127] However, just before the meeting, Numenta had a series of breakthroughs with its mathematical model, and Hawkins was feeling optimistic.

At the heart of Numenta's work is a technology it calls HTM, or hierarchical temporal memory. It can run on a laptop, so it's not relying on supercomputers or new kinds of brainlike hardware. HTM is an attempt to get at the fuzzy logic of the brain, especially its ability to connect lots of different, noisy bits of information to a concept. As discussed earlier, a young brain might connect everything that has to do with a frog, so that later the sound of a croaking frog or a drawing of an outline of a frog or seeing just part of a frog will connect back to "frog" and open up that concept. Such a feat would prove nearly impossible for a computer, which can't learn the generalized concept of a frog. It could be fed millions of images of frogs, but if then it came across an image of a cloud formation that looked remarkably like a frog, it wouldn't have the slightest idea how to make the connection to "frog." All it would see is a cloud.

Hawkins's team is trying to get computers to build hierarchical layers, like the brain does. The lowest level would gather vast amounts of input from all kinds of sources—the equivalent of a person absorbing raw information from all senses. That raw data would be passed up to a next level, which would start to assemble the information into concepts, and up to another level that would look for associations and assemble a richer concept, and so on up the line. At the same time, the highest levels in HTM's hierarchy would generate expectations about what it's looking for and compare those expectations to the information coming up from lower levels.

Part of the challenge of HTM is getting it to "learn" about something. If HTM is going to learn to look for frogs, it has to be fed lots of images of frogs—with one key additional step. The images are presented to the computer over time, and it scans them from top to bottom and then left to right. This helps the computer realize that pieces of a frog are part of a whole frog. Each layer of HTM has to be trained separately, so every layer is loaded with all the information it needs to make an inference. The problem with this kind of learning is that it's slow and individualized. A child takes years to learn enough to find a frog in a picture book. We probably don't want computers that take years to get programmed.

Though the process is tedious, when completed, HTM should be able to "watch" a lot of unrelated, noisy input and figure out what it's seeing. If someone wants an HTM program to watch for frogs in a pond, it could train multiple cameras on the pond and report back anytime it sees any kind of frog or just a frog's head peeking above the surface. One of the first practical applications of Numenta's HTM is in video detection technology from a company called Vitamin D. The technology can, for instance, learn the basic images of the man, woman, and dog living in an apartment. If a security camera is then set up in the apartment and someone comes in who does not live there, the system can recognize the irregularity and generate an alert that an intruder might be in the place.

The input to HTM doesn't have to be visual, and this is where computers with some brainlike smarts can do things humans can't. HTM software could watch all the traffic on a computer network and develop a model of what that traffic looks like when it's normal and what it looks like when a hacker is starting an attack. Then the system could predict when an attack might be coming and generate an alert or take action to stop it. HTM could do this kind of analysis of credit card transactions to look for fraud or health information to predict the onset of a flu outbreak or epidemic. HTM could let millions of weather sensors all over the world constantly feed data to a computer that would assemble the data to "see" global weather as it happens, then make predictions about what will happen next.

"It would visualize weather like you and I see things," Hawkins said.

"The real opportunities, no one is even conceiving of them today," Hawkins said. He wants Numenta to build and market a "prediction engine" that can be taught anything and applied in myriad ways—a platform that gets developed by others for innovative uses. "The machine can learn—you don't encode it with software," Hawkins said. "It's a new way of thinking about data."

Numenta, Hawkins said, is only beginning to figure out how to make a commercial prediction engine. "Machine learning today is where computers were in the 1950s: it's really hard to do and can only be done by experts," he said.[128] Numenta's goal is to make HTM into something almost anyone can use. As Hawkins works toward that, he's slowly figuring out more about how the human cortex works and how to put it in a software package.

Meanwhile, the award for most famous computer brain of the early 2010s will likely go to Watson, the IBM machine that defeated humans on the game show *Jeopardy!*

Watson does a remarkable job of understanding a tricky question and finding the one right answer. IBM's scientists have been quick to say that Watson does not actually think. "The goal is not to model the human brain," said David Ferrucci, who spent the fifteen years leading up to Watson's TV stardom working at IBM Research on natural language problems and finding answers amid unstructured information. "The goal is to build a computer that can be more effective in understanding and interacting in natural language, but not necessarily the way a human does it."[129]

Watson's success comes from exploiting a trait of the human brain: it processes many possible answers at the same time, weighs the probability of each of those answers being right, and then settles on the most likely right answer.[130]

Computers have never been good at finding answers. On the old search site Ask Jeeves, you could ask a plain-language question, but Jeeves wouldn't deliver an answer. He'd cough up tens of thousands of search results. University researchers and company engineers have since worked on question-answering software, but the very best can only comprehend and answer simple, straightforward questions (How many Grammies did the Beatles win?) and will still get them wrong nearly one-third of the time.[131] That isn't good enough to be useful, much less to beat *Jeopardy!* champions. The questions on the game show are full of subtlety, puns, and wordplay—the sorts of things that delight humans but choke computers. "What is black death of a salesman?" is the response to the *Jeopardy!* clue "Colorful 14th-century plague that became a hit play by Arthur Miller." The only way to get to that answer is to put together pieces of information from various sources, because the exact answer is not likely to be written anywhere. The computer has to be, in a sense, creative. Ferrucci's breakthrough was to get a massive computer to act just a little like a brain.

Watson runs on a cluster of Power 750 computers—ten racks holding 2,880 processing cores. It's a room lined with black computer cabinets plus storage systems that can hold the equivalent of about one million books' worth of information. Over a period of years, the Watson system—the computer hardware plus software called DeepQA created by Ferrucci's team—was fed mountains of information, including text from tens of thousands of books, all of Wikipedia, millions of news stories, rhyming dictionaries, synonym finders, and more. The IBM team stopped adding to the data stored in Watson only when it became clear that additional information was no longer improving the results.

When a question is put to Watson, more than one hundred algorithms analyze the question in different ways and find different plausible answers—all at the same time. Yet another set of algorithms ranks the answers and gives them a score. For each possible answer, Watson finds evidence that may support or refute that answer. The answer with the best evidence assessment earns the most confidence. The highest-ranking answer becomes *the* answer, though during a *Jeopardy!* game, if the highest-ranking answer isn't rated high enough to give Watson confidence, Watson might decide not to buzz in and risk losing money if it's wrong. The Watson computer does all of this in about three seconds. In early 2011 Watson was good enough at finding right answers to play on TV against former *Jeopardy!* champions Ken Jennings and Brad Rutter. Watson's victory made headlines worldwide—a cultural milestone. While answering the last Final Jeopardy clue, Jennings wrote on his screen the rueful comment, "I, for one, welcome our new computer overlords." Yet compared to the way human brains work, Watson goes through a woefully inefficient process that relies on brute force and a whole lot of electrical power. Searching billions of pieces of information and finding hundreds of possible answers to get one right answer is kind of like razing a rain forest to find a single tree. Still, Watson is a step in a very difficult progression toward brainlike computers that can tackle tasks that remain out of reach of traditional computing. Watson's question-answering technology could evolve to handle online medical diagnosis. Instead of searching endless Web pages about a health condition, a person could connect to a site powered by question-answering software and type in a sentence describing symptoms. The software would analyze the symptoms against all current medical literature and all available information about the patient to come up with the most probable diagnosis—something like the process that would happen in a doctor's brain. Imagine that replicating across the commercial landscape. A shopper could tell an online Watson-driven service her size, her price range, and the exact kind of white shirt she was looking for, and Watson would come back with the best choice instead of a search page packed with every item tagged with "white" and "shirt."

"I want to create something that I can take into every retail industry, in the transportation industry, you name it," John Kelly, who runs IBM Research, told the *New York Times*. "Anyplace where time is critical and you need to get advanced state-of-the-art information to the front decision-makers. Computers need to go from just being back-office calculating machines to improving the intelligence of people making decisions."[132]

IBM's Watson and Numenta's HTM aren't, by themselves, two-second-advantage technologies. They're not providing the kind of tailored, practical, real-time, event-driven capabilities that are being deployed by companies such as Sam's Club and Caesars Entertainment. But IBM and Numenta are working to emulate brain processes in software, and that kind of research will help inform and drive coming generations of two-second-advantage technologies. The more computers borrow from brains, the better they can get at learning from experiences. To get really good at learning, hardware and software and memory and processing are all going to have to become the same thing, as it is in the brain—and the technology is a long way from that. For now, there are only these piecemeal advances at the edges—experiments such as memristors and neuromorphic chips in hardware and HTM and Watson in software.

Still, the research is helping in the development of more brainlike computers. The capability is emerging to make computers that can build efficient models of how things work, absorb real-time events, and predict what will happen next. Advances in hardware and software provide a real path out of the von Neumann bottleneck and toward talented machines.

If there's a center of gravity for automated intelligence, it's in the financial markets. That's a good place to see the real, live contrast between computer brains and human brains.

TIBCO played a significant role in introducing trading floors to real-time information. In the early 1980s, government regulation of traders eased, discount brokerages slashed commissions, and members of the public raced into the stock market. By the middle of the decade, the volume on the New York Stock Exchange was around one hundred million trades per day—twenty times the volume of the 1960s. But technology hadn't kept up, and trading houses were swamped by the sheer avalanche of trades. Each trader had a dozen or more screens on his desk, all connected to different data sources. On average, it took over twenty keyboard entries to record the details of a single securities purchase—a far cry from a click of a mouse, much less an automatically triggered trade.[133]

TIBCO developed a software bus for Wall Street—a technology translator to which any component or application could be connected. It allowed various data sources to flow into one computer, where the different bits of information could be collected, translated into a common software code, and presented to traders on a single screen. In 1987 TIBCO built a single-screen system for Fidelity Investments, blending up to twenty-five different news reporting sources, prices, charts, company financials, and so on. Soon much of the financial industry adopted information bus technology, and the real-time era was born. Public information about stocks and bonds became available instantly to any computer. If you think about it in human terms, financial firms' computers were granted an ability to "see" all of what was happening in the global financial markets through data, at a speed and intensity that humans could never comprehend.

But while this sensing ability was given to computers, the computers weren't processing and "thinking" about what they saw. That was still being done by the humans—the traders.

Enter the mathematicians and software engineers. They were recruited by financiers to write increasingly sophisticated programs that watched data streams and took action based on sets of rules. The programs could watch more data and react more quickly than humans. Software could spot tiny stock fluctuations and trade millions of shares in a split second. Complex algorithms could churn data through ever-faster computers, taking on, in a way, a life of their own. By the mid-2000s, algorithmic trading had overtaken the financial industry. It generated vast amounts of wealth—that's how hedge funds made billions of dollars a year—yet also showed its vulnerabilities as software made bad decisions at crucial times, amplifying the financial crisis of the late 2000s.

Trading systems had developed a kind of intelligence, but not a human intelligence. If an event happened—a company's earnings report, a terrorist bombing, a recall of a popular product—the systems would see it through data, use their rules to make predictions about what might happen, and make trades or take other action. "The kind of trading strategies our system uses are not the kind of strategies humans use," Michael Kharitonov, a computer scientist and CEO of Voleon Capital

Management of Berkeley, California, told *Wired* in 2011. "We're not competing with humans, because when you're trading thousands of stocks simultaneously, trying to capture very, very small changes, the human brain is just not good at that. We're playing on a different field, trying to exploit effects that are too complex for the human brain."[134]

In the financial markets, computer intelligence is doing what it does best: complexity and math. Those happen to be two things humans are bad at. The computers and humans act as partners. In the financial markets, though, for the first time there is a question of which intelligence is in charge: the computers or the humans? *Wired* concluded: "It's the machines' market now; we just trade in it."[135]

And yet there are key capabilities that totally escape machines. One is that they can't learn in any meaningful way. All of the sophistication of the machines comes from the humans who write the algorithms and software. The machines can't improve themselves or decide to go in a different direction on their own. They can't use all that incoming data to significantly better their mental models, the way Mayor Menino did with his model of Boston. They can't rewire themselves based on experiences to generate ever-faster and more accurate predictions, the way Wayne Gretzky did in his years of playing hockey. It is the machines' fundamental weakness—the chief reason we are still their masters (despite Ken Jennings's comment) and they are still our tools.

The neuroscientists and computer scientists are trying to change this and make computers that can learn—well beyond the limited learning of forward-chaining systems. It is, ultimately, what everyone mentioned in this chapter is working toward, and it would be the biggest leap in computing since pioneers like Grace Hopper and John Backus invented programming in the 1940s and 1950s.

Beyond learning, machines have another handicap. Psychologist George Miller put it best: "The crowning intellectual accomplishment of the brain is the real world."[136] Everything that we perceive as the real world is not particularly "real" at all. It's a collection of crazy quantum phenomena. Objects, motion, heat, light, dogs, cats, bacon, Merlot, footballs—they're all a mixture of atomic particles and physics that our minds assemble and assign properties to. They're not absolutes. "One theory is that the brain creates a model of the universe and projects this version like a bubble around us," said Henry Markram, who's building the Swiss brain simulation. "Ninety-nine percent of what you see is not what comes in through the eyes—it's what you infer."[137] Computers can't see a bookshelf. They can be fed data that describes a bookshelf and perhaps even understand a bookshelf at a level of detail humans could not register. But computers don't sense a bookshelf and know what it feels like and how, for instance, it keeps books suspended in place instead of letting them melt through and fall to the floor. As with financial markets, we can add sensors and generate data streams about any aspect of the real world, but computers will still see that thing as data. Humans see that thing as a thing.

A key aspect of the real world is other humans, and computers will never understand us. Humans have a unique capability called the theory of mind. We understand what other people are doing because we model and guess at what's going on in their heads. If you see a man walk up to a woman and kiss her, you make some assumptions about what's going on—the two are a couple, the man is glad to see her, and so on. You don't look at it in purely physical terms; otherwise all you'd see is that a man ran into a woman and their lips smashed together. "Our minds explain other people's behavior by their beliefs and desires because other people's behavior is in fact caused by their beliefs and desires," wrote psychologist Steven Pinker. "We mortals can't read other people's minds directly. But we make good guesses from what they say, what we read between the lines, what they show in their face and eyes, and what best explains their behavior. It is our species' most remarkable talent."[138]

It's impossible to understand and function well in the humans' real world without having a theory of mind. In fact, this is the difficulty facing people with severe autism. They can't model other people's minds—can't understand their intentions, actions, feelings. It leaves autistic people stranded inside their own minds. Like Stephen Wiltshire, the savant artist from earlier in the book, an autistic person may have stellar brain hardware and even software capable of outstanding feats, but without a theory of mind, the person is severely handicapped when dealing with the broader world.

Unless computers actually have a human mind, they're not likely to ever model the human mind. Let's say the scientists manage to build a computer that can learn and rewire itself and build efficient models and make excellent predictions, getting very close to copying the modus operandi of the human brain. The computer still won't *have* a human brain and so won't understand humans or the human world. The computer might far exceed human capabilities, as financial trading computers do. But basically, it will be autistic. We will wind up making autistic savant computers.

Futurist Ray Kurzweil and others describe a "singularity"—a point in time when the world's connected computers collectively become smarter than humans and, presumably, take over.

"There will be no distinction, post-Singularity, between human and machine," Kurzweil wrote.[139] His predicted date of the singularity is 2045, when he expects computer-based intelligences to significantly exceed the sum total of human brainpower. It's the Terminator story with a dash more plausibility. Yet even the singularity misses the theory of mind. Computers might someday create their own version of reality, and maybe that version could threaten human reality. But they won't take over our reality unless, somehow, the machines pull a Pinocchio and actually become human.

Instead of feeling anxious about brainlike machines, it would be wise for society to embrace them. Such machines are still our tools, even on Wall Street, where the humans can write new rules and regulations and make the machines conform to them by altering their hardware and software. The machines are beginning to help police prevent crime, give companies a way to make customers happier, save the lives of premature babies, and ease traffic in cities. The case studies we've written about are only the beginning. A computer winning on *Jeopardy!* is basically a gee-whiz moment on the way to something bigger. Computers armed with just a little predictive talent can already solve interesting problems and change the way organizations work. In the not-distant future, talented machines can become supercapable servants, helping to fix some of the world's great problems. The two-second advantage is much more than a passing technology trend. It's a major leap in the evolution of machines and their relationship to the human race.

PART III

THE TWO-SECOND ADVANTAGE

THE TWO-SECOND ADVANTAGE AND A BETTER WORLD

On October 24, 1907, a carriage pulled by a white horse nosed through a throng that had gathered at the corner of Wall and Broad streets in New York. It stopped in front of the granite offices of J.P. Morgan & Co., and the intense, bulbous-nosed financier J. P. Morgan stepped from the carriage and into the building as the crowd strained to get a glimpse of him. The U.S. financial markets had been falling apart for days. Spooked by a failed attempt by two men to corner the copper market, jittery investors had panicked and sent the New York Stock Exchange into a free fall. That set a spiral in motion, as institutions and people pulled money from banks, which caused bank failures, which in turn scared more people into pulling out their money. By the time Morgan got to his office that day, it was packed with men asking him to do something to save the financial industry.[140]

This was an era before the United States had a strong central bank. The Federal Reserve had yet to be created. It was left to Morgan, a private banker, to take charge. He promised to inject millions of dollars of his own money into the system as loans. He got the U.S. Treasury secretary to offer $25 million in additional liquidity and convinced some of the nation's wealthiest barons, including John D. Rockefeller, to deposit millions more into banks. By midafternoon, the stock exchange was threatening to close early to try to stop a complete free fall of stock prices and the sudden collapse of brokerage firms. Morgan insisted the market stay open until its usual closing time of 3:00 p.m. When the market managed to stay open until the final bell at three, a roar of yelling and cheering could be heard coming from the trading floor. Morgan had prevented a total meltdown, although it wasn't a complete victory. The economy fell into a deep recession in 1908.

The panic of 1907 reverberated through the political system over the next few years, leading to the Federal Reserve Act of 1913, which established the Fed as a quasi-government central bank with a duty "to furnish an elastic currency, to afford means of rediscounting commercial paper, to establish a more effective supervision of banking in the United States, and for other purposes," according to the act. For most of the rest of the twentieth century, the Fed performed those tasks with varying degrees of success—disastrously in the 1930s and decently in the final twenty years of the last century.

In autumn 2007, however, the Fed ran into trouble. Almost exactly one hundred years after Morgan's heroics, the worst crisis in a generation tore through the U.S. financial markets and severely damaged markets worldwide. It was set off by the bursting of the U.S. real estate bubble. Its reverberations collapsed risky positions held by financial institutions, caused a liquidity crisis, and forced massive government bailouts to keep the financial industry from completely falling apart. As this book was being written in 2010 and 2011, world economies were still recovering from this mess.

One of the astounding things about the massive systemic failure of 2007 is that the Federal Reserve didn't see it coming—at least not clearly enough to preemptively steer the country away from disaster. The Fed and its chairman, Ben Bernanke, have since been criticized by politicians and pundits, some even suggesting that the Fed should be abolished. Leaving political arguments aside, what's clear is that the Fed *technologically* operates like a twentieth-century entity. It relies on massive amounts of data served up in batches, informing Fed governors about economic conditions in the near past but not the present or near future. The Fed's decision making, by twenty-first-century standards, is slow and reactive, like turning the wheel of a car after it hits a pothole. After the panic of 1907, the world realized it needed a new way to keep an eye on the economy because the old way clearly no longer worked—the economy had gotten too complicated for the system that was in place. After the crisis of 2007, the same is true again. The world needs a new way to watch and make adjustments to the economy.

Some might say that the solution is to pass legislation and eliminate or overhaul the Fed. But here's perhaps a more practical idea: give the Fed two-second-advantage technology.

Briefly put, the Fed's job is to ensure a suitable temperature in the economy so that inflation and employment stay within an acceptable range. (We want some inflation but not too much and some unemployment but not too much.) How does the Fed do that? Mostly, it meets eight times a year to decide whether to increase or lower interest rates. It never changes rates by less than one quarter of a percentage point (also known as twenty-five basis points). Raising rates tends to pull money out of the economy and slow things down, while lowering rates frees up money and generally stimulates the economy.

How does the Fed get input about the state of the economy so the governors can make such critical decisions about interest rates? Six weeks before each Fed meeting, the regional Fed banks survey a range of businesspeople and bankers in their regions. They also pull together numbers about economic conditions in the area. They assemble all that information into a report called the Beige Book that goes to the Fed governors. By the time that information gets to the Fed, it's old news.[141] It describes what's already happened, weeks or months prior. The same is true of most all of the information that might influence the Fed: consumer confidence figures, unemployment rates, factory bookings, home sales, etc. All those reports contain information about the past. About the only real-time information the Fed sees are the numbers on the stock market ticker.

If you applied the Federal Reserve approach to ensuring a suitable temperature in your home, you would turn the heater on and off eight times a year, usually overheating or underheating your house. Since the late 1990s, that hasn't worked too well for the Fed. The United States has gone through a wild dot-com boom, then the dot-com crash, then an equally crazy housing boom, and an even worse financial meltdown ignited by the housing bust.

So, then, how might the Fed work better in an era of two-second-advantage technology? Start with the algorithmic trading systems discussed in the previous chapter. Those systems continuously watch real-time data streams from a number of sources—prices of stocks, bonds, commodities, and currencies, but also news from sources such as Bloomberg and Reuters. They run that data through models that make predictions about where certain investments are heading, and they take action in the form of trades, all without human intervention. If private systems can watch those data streams and generate instant predictions about where things are heading, the Fed can too. In fact, the Fed could be given the right to go even more meta and tap into the different private algorithmic trading systems—so the Fed could see what they were seeing and what they were doing.

The Fed could, in addition, have access to data streams that Wall Street systems don't see. We wrote earlier about Sam's Club tracking purchases of individual members and being able to predict with great accuracy what they're going to buy next in a certain window of time. The Fed could tap into that, in real time, and always have a picture of what consumers were about to buy—instead of always looking at what they bought weeks or months ago. Xcel Energy and other utilities are increasingly putting in smart systems that take real-time readings and predict energy use, which can be an economic indicator. Federal Express relies on a sophisticated real-time system that always knows how many packages are moving to which cities, which could provide a constant barometer of regional economies. Intel, Ford, Wal-Mart, Dell, and many other companies employ systems that deliver real-time sales and manufacturing data. As the next decade unfolds, much of the economy will be operating on real-time and predictive systems.

The Fed could watch all of it—constantly. It could see the U.S. economy the way Wayne Gretzky sees a hockey game in progress: thousands of factors all instantly captured and processed and understood. And like Gretzky, the Fed's systems could see patterns and make predictions. But, importantly, if the Fed is going to govern the economy the way Gretzky plays hockey, the Fed can't meet eight times a year and try to predict what's going to happen to the economy over the coming months. And it can't adjust rates by twenty-five basis points or more, which, in interest-rate terms, is quite a big movement. Instead, the Federal Reserve would have to hand over the power to adjust rates to its predictive system, which would adjust rates *all the time.* Instead of trying to guide the economy with long, sweeping actions based on long-term guesses about economic conditions, the Fed would steer the economy by constantly making very short-term, highly accurate predictions about economic conditions and adjusting rates on the fly by as little as one hundredth of a percentage point.

This doesn't mean completely turning over economic policy to machines. It means, in fact, that the Fed governors would be freed to concentrate on overall economic policy and leave the dirty business of rate adjusting to the machines. The governors would decide the rules the machines would play by. How much inflation do we want? What kind of housing market? How much speculation in the markets? And certainly the governors could override the machines, pulling the plug if something went awry or taking extreme action in an emergency.

The system would be transparent. No more reading tea leaves to guess whether the federal funds rate was going to head up or down at the next Fed meeting. Rates would constantly change, and anyone should be able to see those changes and follow them on a computer screen, just as investors can follow the Dow Jones Industrial Average or the yield on Treasury bills.

The Fed could collect and see real-time data from a variety

of sources—Wall Street firms, retailers, manufacturers, transportation companies—using technology available today. That would be a start—at least when the Fed met, it could see a picture of what was happening right then, not six weeks ago. Writing the algorithms and building the model that could tell such a system how to see the whole economy on its own, make predictions, and take action would be difficult and take time. But if math jockeys can do that kind of thing for hedge funds, they could do a more complex version for the Federal Reserve.

Building a system that could truly watch the economy and learn what it does and from that make ever more accurate predictions—this is the stuff still going on in labs. It probably requires post–von Neumann computing: parallel processing, adaptive so it can wire itself (like neuromorphic chips), hierarchical software (like Numenta's HTM). But it's not the stuff of dreams. It's stuff that's going to be real in a decade or so.

The biggest barrier of all would probably be the very idea of ceding to machines what appears to be the Fed's main function. But monkeying with interest rates eight times a year is *not* the Fed's main function. As the Federal Reserve Act said, the Fed is there "to furnish an elastic currency, to afford means of rediscounting commercial paper, to establish a more effective supervision of banking in the United States, and for other purposes." Interest rate adjustments are a tool—albeit an important one—that the Fed can use to do its real job of making sure the economy functions smoothly and effectively. Machines that constantly read the economy and adjust rates will be able to automate that tool and wield it more efficiently. And machines will be able to keep up with the machines already deployed by financial firms. Hopefully the Fed's machines would then have a better chance of preventing Wall Street's machines from leading us into another boom-and-bust cycle.

Two-second-advantage technology is arriving. These event-driven systems can form models by analyzing massive amounts of data, but they don't rely on accessing that data all the time. Borrowing from the way the human brain works, these systems are predictive—they take in real-time events, predict what's about to happen, and take action or send a notification without human intervention. They operate on the idea that a little bit of the right information ahead of time is more valuable than piles of information too late.

As technology allows computers to work more like the prediction engine in the human neocortex, computer systems will get more and more "talented"—maybe not exactly like human talent, but talented in the sense of being able to make predictions better and faster than competing systems.

The trend toward talented systems will continue and gain strength—because it has to. Computer systems that rely on twentieth-century technology—on von Neumann architecture—increasingly can't keep up with the floods of data and instantaneous pace of the 2010s and beyond.

Smart leaders are planning for this transition. They realize that two-second-advantage technology can drive a rethinking of the way they run enterprises and do work. This will be an important shift over the next decade. Organizations that don't take advantage of the leap to brainy, predictive technology will be rendered as outmoded as a silent film company after the arrival of talkies.

In the more immediate future, two-second-advantage technology could help solve some critical information problems. One of those is the global fight against terrorism.

Early in the book, we recounted the attempted bombing by Umar Farouk Abdulmutallab of Northwest Flight 253 on Christmas Day 2009. Numerous pieces of information that might've led authorities to prevent Abdulmutallab from getting on the flight were stuck in separate computer systems, unable to be assembled and understood by any one entity. Abdulmutallab had overstayed his visa in Yemen (to attend an al Qaeda training session). His father had warned the U.S. embassy in Nigeria about his son. Abdulmutallab had paid $2,831 in cash for his airline ticket to Detroit, where he had no known relatives. British authorities had already denied Abdulmutallab a visa. A number of data points were floating around the world. Individually, each had little meaning. Together, they suggested a potentially dangerous radical.

A similar story could be told about the May 1, 2010, attempt by Faisal Shahzad to set off a car bomb in the middle of New York's Times Square. The city has eighty-two surveillance cameras pointed at Times Square, and several could have glimpsed a suspiciously moving Pathfinder as it entered the tourist-jammed area and parked in a no-parking zone. But if nobody was watching the particular camera that caught the vehicle—which, apparently, was the case—the images from the cameras went into a void. Shahzad, who had lived with his wife and child in the United States, had earlier in the year gone to Pakistan for five months with his family, then returned to the United States alone—a bit of information that might have added to reasons to watch Shahzad. The Pathfinder he was driving had license plates that he'd acquired illegally and wouldn't have matched the vehicle. He had made calls from a mobile phone to a number that was in a Homeland Security database of numbers potentially associated with terrorism. All of these bits of data remained separate. They added up to nothing—until the bombing failed and human investigators began piecing information together.

Antiterrorism entities could work differently if they employed two-second-advantage technology. Data from dozens or hundreds of different sources could stream into the same system. Analysts could work with programmers and mathematicians to develop rules about what to watch for in the data streams and how to build models of every person of suspicion—not unlike Sam's Club building models of every card-carrying member, except that the suspected terrorists don't get to opt out. As the system built its model, it would start seeing patterns that would allow it to make predictions. If some guy spent time in Yemen and then got another suspicious ping—like a father's warning to an embassy official—and then bought a plane ticket to the United States with cash, the model might correlate that activity to someone highly likely to do something violent and refuse him a seat on the plane.

Maybe each of those activities by themselves wouldn't mean much, and even a human analyst might never spot the way certain activities point to certain predictions. But it's not all that different from knowing that if a Sam's Club member buys diapers, a coffee machine, and a package of steaks, he's likely to buy a big-screen TV in the next month. Technology can do this today, and it will get far more sophisticated in coming years.

Two-second-advantage technology could also help avert a different kind of life-threatening crisis: a flu pandemic. The strategy behind stopping an outbreak of a dangerous, highly contagious flu from becoming a global pandemic is much like that of stopping a forest fire from spreading. Officials first have to spot the outbreak, then cut it off by getting out in front of where it's heading and vaccinating everyone. Outbreaks are going to happen, but if they can be contained, they'll be far less destructive. As with forest fires, the earlier a flu outbreak is identified, the easier it is to contain.

Modern communications have certainly made flu tracking much better than ever before. Doctors around the world report flu cases to the World Health Organization, which by 2009 had begun posting confirmed cases on a map on the Web for anyone to see. But doctors' reports filtering through a WHO bureaucracy are not exactly real-time or predictive information. They're a glimpse of where the flu has been days or weeks before. By the time an outbreak is noticed in a region, people from that area could have boarded planes and flown all over the world, unaware that they're carrying the virus and transplanting it somewhere else.

In 2009 Google assembled an interesting new way to watch for flu outbreaks in something closer to real time. As Google long ago figured out, search terms correlate to events in a person's life. People who search for terms like "housebreaking puppies" probably just got a dog. People who search for real estate information are probably looking to move. And people who type in terms like "flu" and "vaccine" and "fever" are usually either sick with the flu or caring for someone who has the flu. By sorting and tracking billions of search queries by location, Google is able to create a constantly updated map that shows probable flu hotspots. "We compared our query counts with traditional flu surveillance systems and found that many search queries tend to be popular exactly when flu season is happening," Google's philanthropic arm, Google.org, posted on its blog. "By counting how often we see these search queries, we can estimate how much flu is circulating in different countries and regions around the world."[142] Google published an article about the method in the prestigious scientific journal *Nature* but noted that to accurately report flu activity, the map would lag searches by about a day. And as might be expected, Google's method of flu tracking only works in parts of the world where lots of people are using Google—which means primarily developed countries.[143] Still, Google is on to something—search queries are a real-time source of information about flu activity.

F. Hoffmann-La Roche, the Swiss pharmaceutical giant usually referred to as Roche, has its own ways of watching for flu outbreaks. Roche has a big stake in predicting the location and size of outbreaks because it makes Tamiflu, the antiviral drug most often used to keep outbreaks from spreading and causing deaths. Tamiflu presents a particular challenge. The primary ingredient is shikimic acid, which comes from star anise, a spice found in the star-shaped fruit of a small evergreen tree found almost exclusively in China and harvested between March and May. The manufacture of Tamiflu requires a ten-step process that involves fermentation and takes six to eight months to complete. In other words, there is no way for Roche to suddenly boost production to respond to an unexpected flu epidemic. Roche needs to be predictive. The company built a system on an event-processing platform that tracks flu information from disparate sources around the world, constantly matching incoming data against models of past flu epidemics. The system not only helps Roche figure out how much Tamiflu to make but also where to send existing supplies. If Roche has a cache of Tamiflu in Canada but an outbreak begins in Vietnam, the company can move supplies to head off the spread of the virus.

None of these approaches—the WHO database, Google's search-term watch, Roche's Tamiflu demand predictor—by itself give the world's health officials a two-second advantage over a devastating mutation of the flu virus that could kill thousands or millions of people. Ultimately, we need to detect a pattern a little bit before the flu has a chance to take hold and move out from a small geographic area. WHO could move in that direction today. The key is to collect all the possible events from all the possible sources and watch the world's health the way hedge fund algorithmic trading computers watch the markets. WHO would not only get data from the world's medical professionals but also watch search terms—not just from Google but from search engines and health Web sites all over the world. Such a system could also incorporate streams from companies, such as Roche, that have a stake in predicting flu outbreaks. The more streams from disparate sources the better. Those data streams then would zip through an agile model that would instantly look for patterns, perhaps knowing that if it sees one thing from Google, another from Roche, and another piece of data about how many people visited a certain medical clinic, there's a strong possibility of flu in a certain region.

This kind of predictive global health will get even better as personal health becomes more monitored. Consider what's being developed by Sisters of Mercy Health System, a not-for-profit network of twenty-six hospitals and 1,300 physicians based in the U.S. Midwest. In 2007 Mercy started implementing real-time monitoring of patients in its hospitals and certain categories of patients who were either discharged from hospitals and needed to be watched or had chronic conditions such as diabetes. Like the USCIS person-centric approach described earlier in the book, the Mercy system takes a patient-centric approach. A patient with a chronic disease might wear a wireless device that can take vital signs like pulse rate and blood pressure and constantly send those back to the Mercy system.[144] Information about that patient from other sources—lab results, notices of prescriptions getting filled, notes from personal physicians—gets collected through a software bus and examined by Mercy's software. The system also has the patient's electronic medical record, so it knows the patient's history and can use that to build a model of the patient. The software has a set of rules for each individual patient and keeps an eye on the data about the patient, constantly watching for patterns. "If you're a diabetic and your glucose gets to a certain level and your blood pressure has risen, that might set off a yellow alert," John Conroy, director of application development at Mercy, told us. "In a quarter of a second, if blood pressure goes up or skin temperature changes, it's detected." If the system sees a problem, it sends alerts to appropriate medical professionals. Conroy said the system is identifying an average of two hundred patients a month who need some kind of immediate, life-saving intervention.[145]

On the enterprise level, Sisters of Mercy has developed an early two-second-advantage system. It has put in place state-of-the-art, real-time technology that brings in data from many sources, watches the data through a model of each patient, and tells doctors when a patient is about to have a problem—rather than the alternative, where a doctor ends up reacting to a patient after the patient's condition reaches emergency proportions.

But on a higher level, Mercy is a bellwether for new health-care technology. Hospitals and medical systems are increasingly going to implement this kind of monitoring technology, gathering streams of real-time, intricate data about millions of people. Imagine a global organization like WHO tapping into that information—from Mercy and hundreds of other similar real-time systems. WHO could sort for patterns in individuals that signal influenza. WHO could see signs that individuals were getting the flu before those people even knew they had it. Now imagine WHO flowing that data together with Web searches and corporate predictive systems, learning patterns that come together whenever a dangerous flu is about to hit a population center.

Medical science has not been able to find a way to eradicate or kill viruses—not even the common cold, much less a lethal flu. But two-second-advantage technology can get sophisticated enough to spot a flu outbreak before it even looks like an outbreak. By rushing vaccines to the right place and isolating the area, medical officials could stop the most contagious flu viruses from endangering the world's population. WHO could predict pandemics so they never happen again.

In November 2010 a new ownership group bought the National Basketball Association team based in the San Francisco Bay Area, the Golden State Warriors. The new owners include movie producer Peter Guber, Silicon Valley investor Joe Lacob, real estate investor Erika Glazer, and this book's coauthor, Vivek Ranadivé, who is now the team's vice chairman. As you might imagine, we have some thoughts about what two-second-advantage technology can do for an NBA franchise.

High-level sports programs are pretty aggressive users of technology. Earlier in the book we described how baseball's San Francisco Giants use Sportvision to track and analyze what happens during a game, and we wrote about technology that lets swimming coaches track and analyze the real-time movements of swimmers as they churn down the lane. The MIT Sloan Sports Analytics Conference has turned into a significant event at the Massachusetts Institute of Technology, attended by owners and coaches from every major sport. The conference is all about using information technology to gain an edge in competition. Since the mid-2000s, NBA teams have increasingly used statistical analysis to, for instance, help players know where on the court they shoot from with the highest levels of accuracy. On the business side, teams mine databases to help with marketing of tickets and merchandise.

Almost all of the analytics used in sports in 2011 relies on backward-looking technology. It's based on the twentieth-century idea of using past information to try to predict events hours or weeks or months ahead of time. We see a new generation of technology emerging for sports—one that builds models, takes in real-time events, and makes instantaneous predictions about what's just about to happen. The technology isn't going to create a robotic Gretzky (or, more appropriately for basketball, a Michael Jordan, who undoubtedly had a massively chunked predictive model of the game in his head), but it sure could help coaches and management.

Let's start on the court. In a free-flowing, fast-paced sport like basketball, how can a team use real-time information about events happening during a game? A start-up called Krossover, based in New York City, gives us an idea of what could be coming. It has created technology that lets a team upload video of a game to the Krossover Web site. The company's employees, using a gamelike interface, watch the game and tag every event—each basket, steal, foul, shot, assist, and so on. The tags connect the event with the player, the time on the clock, and the spot on the court where it happened. When the tagged video and database come back to the team, a coach can sort for every shot a certain player missed or look at the percentage of shots made from different places on the court. At this writing, Krossover can only analyze video uploaded to the site after the game, tagged by human employees. CEO Vasu Kulkarni explained that image-recognition technology isn't quite good enough yet to automate the process. But he imagines that such technology is coming—technology that lets computers watch a flowing game and identify distinct events, knowing the difference between a basket and a miss, in real time. The goal for Krossover-like services would be to have technology that could watch the games as they happen and tag events on the fly.

Building on those events plus past statistics that teams have stored in their databases, a two-second-advantage system could start to chunk patterns about the team—and do the same about opposing teams—creating models that could take in events and make instant predictions.

How could that be helpful? Experienced NBA coaches have seen enough games over enough years to already have chunked predictive models in their heads, but they're busy during a game, processing a ton of information about their players, time management, the referees, and strategies. So let's say an assistant coach is carrying an iPad wirelessly connected to the team's two-second-advantage system. The system has chunked patterns about the opposing team and has been tracking the current game and notices a pattern emerging toward the end of the game: with time running out and that team down by two points and facing a particular kind of defense, their star point guard tends to drift to the right, look for a pass in his favorite spot just beyond the three-point line, and shoot to try to win the game. In other words, the system could make a prediction of what was about to happen in the game based on real-time events—and flag this to the assistant coach who's looking at the iPad. The assistant coach might then tell the coach what the system has reported. The team could call a time-out and gather the players around, telling them what to expect and instructing them to jam up that spot on the court and force a different outcome. It's a great example of how a little bit of the right information just ahead of time can be more valuable than a lot of information too late. During a game, the coach would be burdened by a detailed report that told him all the tendencies of the other team's star player. What he really needs to know is what the player is going to do next, in that game, so his team can try to counter it.

On another front, every year NBA owners go through a familiar ritual: the draft. They take turns claiming the rights to young players from around the world, and the draft determines a team's future more than perhaps anything else. The stakes are immense. Owners would be foolish not to employ any and every technology that would help them draft well. They use analytics to sort and rank potential draftees by all kinds of measurements. They model what-if scenarios before the draft, looking at who will be available and what choices to make depending on who gets picked in earlier rounds. Owners come to the draft prepared. But then, as events unfold on draft day, they're mostly left to rely on the models they brought with them. Two-second-advantage technology would handle the draft differently. It would build models based on past experiences in the draft and then be ready to respond to real-time events on draft day, making predictions about what was likely to happen based on what was happening right at the time. It would give team owners a little edge in the important process of picking future players.

On the business side, the two-second advantage will bring some of the benefits that already help companies such as Harrah's and Sam's Club. To a basketball team, a season ticket holder is like someone who signs up for a rewards program—a dedicated fan who is happy to sign up for personalized contact. So give season ticket holders a loyalty card, or build a system that lets each game's ticket become a loyalty card that the ticket holder would scan when making any kind of purchase at a game—or between games, on the team's Web site. After collecting data about season ticket holders' actions for a while, the system would build a model of each individual ticket holder. By noticing the sizes of Warriors apparel that season ticket holder Martha buys, the system might conclude that she has a toddler—who is going to grow. So six months later, she might get an offer for a discount on an item that's a size bigger than the one she previously bought. Or season ticket holder Joe might usually buy a slice of pizza in the fourth quarter, so he's apparently hungry by then. During a particular game, the pizza is almost sold out, but the team is looking at an overstock of hot dogs. As he sits in his seat at the end of the third quarter, the system could send Joe a text message offering hot dogs at half price. It might make Joe happy and help get rid of the extra hot dogs.

As the system gets more predictive, it could be more helpful. It could start to see patterns in season ticket holders who become unhappy and don't re-up the next year. Maybe they increasingly give or sell their tickets to other people—the system would notice because those other people would buy different food or souvenirs than the season ticket holder bought. Just as an unhappy season ticket holder is about to drop out, she might get a call from her favorite player (which the system would know because she always buys shirts with his number) asking her to stick around. As predictive technology develops, sports managers will use it in imaginative ways to make fans happy.

Gaining a two-second advantage in sports may not rise to the level of helping the Fed guide the economy or fighting terrorism or avoiding pandemics. And yet we started this book with hockey star Wayne Gretzky as a way to show the fantastic value in predicting the future a little faster and better than anyone else. Sometimes, sports help us see what's possible, which can be enough to change the world.

As far back as the early 2000s, Eric Schmidt was thinking about how a next-generation Google would get predictive and personal. The Google we all know and probably have mixed feelings about is twentieth-century technology. It knows very little about you and has no idea what you're looking for or why. It sits there waiting for you to ask something of it, and when you do, it delivers millions of results ranked by an algorithm that determines how important each link is to the world at large—but not necessarily to you. Type in "turtles" in early 2011, and you get 15,800,000 results ranging from images of turtles to the Web site for the 1960s rock band the Turtles to the site for the Teenage Mutant Ninja Turtles. Not exactly targeted.

In 2004 Schmidt—then CEO of Google but now executive chairman—first talked to us about his vision for Google 2.0. "I keep asking for a product I'd call Serendipity," he said back then. This product would, like Google, have access to the world's digital information, but it would also look at everything the user had ever worked on and saved to his or her personal hard drive and build a model of the user's work, tastes, friends, and predilections. "Then when I'm typing a paper, it would know what I'm writing about and say, 'Hey, you forgot this,'" Schmidt said. In other words, this Serendipity thing would watch what you're doing in real time and predict what you might need. It would, in short, get to know you and anticipate your needs. It would be an electronic Radar O'Reilly.[146]

Six years later, in the fall of 2010, Schmidt was still hoping for what he by then was calling a "serendipity engine." He talked about it at a conference, saying it would be one of many "new services that just make your life work."[147] Even though it's taking quite a while for Schmidt's vision to become a reality, and it raises yet more privacy concerns, it's still a viable vision. Next-generation search has to go the way Schmidt described. It will build a model of you—assuming you allow it—and work like the brain of a really good personal assistant. It would no longer deliver to you all the world's information after you type a request—it would deliver the information you want just before you need it. If Google doesn't do this, some newcomer will. Or perhaps IBM's DeepQA will evolve into such a service. The way we work and live our lives is going to be transformed by real-time predictive technology.

One of the best-known examples of predictive personal technology—and a good way to see the state of the art—is the Netflix recommendation engine. If you're a Netflix customer, the system records what movies you order and how you rate the movies you watch. It builds a model of your movie tastes and matches you with other people who have a similar model. There's a good chance that movies those people like will be movies you like, so Netflix recommends them to you. The more accurate the recommendations, the happier Netflix is able to make its customers. It's a narrow, movie-focused version of Schmidt's serendipity engine, anticipating what you might want before you even ask for it. Netflix takes this so seriously, in 2006 it launched a global contest to improve its recommendation engine. Mathematicians and computer scientists from all over the world teamed up to enter. In 2009 Netflix CEO Reed Hastings awarded the one-million-dollar prize to a team called BellKor's Pragmatic Chaos. How much did they improve the recommendations? By only about 10 percent. That gives you an idea how hard it still is to get computers to work like brains.

Still, the Netflix contest was widely followed because the technology has implications well beyond movie choices. Just consider what it could mean in the broader realm of personal entertainment. The idea would be to let a single Netflix-like engine loose on your entire realm of entertainment. It should get to know not only what movies you like but also your preferences in TV shows, music, sports, Web sites, and books and build a model of your tastes. You might never have to program a DVR again(assuming we have DVRs in a coming era of on-demand video)—the DVR will be able to predict what you'd like and record it for you. The system should be able to cross-reference, knowing what music you've been listening to lately and alerting you that a band that fits your taste is playing an upcoming gig at a local club. It should be able to continually adjust to events, so at any given moment it could suggest a handful of things you might want to do, taking into account not just your tastes but time, day, weather, and other factors. Certainly humans can make plans for themselves, but the value—as with the Netflix engine—would be giving you ideas you might not have thought of. A computer could tirelessly scan all possible entertainment and only offer up what it believed you'd like, saving you all that work.

As the serendipity engine and Netflix entertainment engine show, two-second-advantage technology for consumers is not likely to reside in a single laptop or other gadget—it's going to be in the cloud, so to speak, out there on servers the consumers will never physically touch.

Patrick Grady is a longtime believer in the concept of predictive personal technology, and he's actually built it into a company called Rearden Commerce. You might not be aware of Rearden because its technology runs anonymously behind a lot of corporate travel services, but the company has raised more than $220 million in financing from the likes of American Express, JPMorgan Chase, and Foundation Capital, and the Rearden Commerce Platform is used by around three million people who travel for companies.

Rearden for now is focused on travel; the technology is supposed to work like a personal assistant who gets to know you. It builds a model of your travel likes and dislikes, such as whether you prefer to stay in a hotel in the center of town, eat sushi, and take the first flight out in the morning. It also can know that you like to see Cubs baseball games whenever you're in Chicago, so if you book a trip there, the system will check to see if there's a game you can catch and offer tickets. Again, this is where the two-second advantage is heading on a personal level. The current practice of personally searching Web sites to book flights and find hotels and make plans is so twentieth century, and too much work. A service built on two-second-advantage technology will anticipate what you need once you tell it that you have to fly to Chicago on Tuesday morning and back Thursday night. It will work from a model of you and figure in real-time events like weather and hotel deals. It might be wrong some of the time, but like the Netflix movie engine, it will learn as it watches what you accept and reject. Rearden has built an early version of this capability, and the technology will only march forward.

Personal health care will change as well. Sisters of Mercy is doing industrial-strength medical monitoring of seriously ill patients, but the technology behind it will get cheaper, smaller, and so easy to use that even healthy people can get monitored. Already devices like Fitbit are available. Fitbit is barely bigger than a quarter and costs ninety-nine dollars. Inside is a sensor that tracks your movements and wirelessly sends information to a computer about calories burned and levels of activity. Such a tiny device could be made to monitor temperature, heartbeat, blood pressure, oxygen levels, and other factors. Someone who chose to wear one could wirelessly transmit a stream of real-time health events to a predictive system. Like the Sisters of Mercy system, a personal health model could draw from past medical records, family history, prescription drug information, and other health events to build a model of you. Real-time data from the personal health device would constantly stream through the model, and the model would learn your patterns and watch for problems. Talk about predictive—such a system would be able to tell you that you were about to get sick before you even felt bad.

Importantly, the health system would then have a rich model of you—it would have chunked you. Because it would be watching your complete health picture all the time, it would know you better than any physician could. Now imagine overlaying question-answering technology like IBM's *Jeopardy!*-playing system. Let's say you're having a health problem—a strange rash, vertigo, or something else a vital-signs monitor might not pick up. You log in and start asking plain-language questions about your condition. The system could instantly search for answers in all the medical literature, compare them to the model of your health, and factor in other real-time information such as whether a strain of flu has rolled into your town. The system could run through all of this and make an initial diagnosis—a two-second-advantage prediction about your condition.

Such a system could take the place of millions of initial doctor's office visits—the kind that lead a family practitioner to do little more than tell you to go see a dermatologist. On the flip side, it could get millions of people medical care earlier, preventing conditions from getting worse and requiring hospitalization or long-term care. As the world worries about how to handle exploding health-care costs, two-second-advantage technology promises to help. "Predictive treatment is driving down the overall cost of providing health care," said Jeff Bell, Sisters of Mercy's chief information officer.[148]

Finally, how about your social life? Think about how static Facebook is. You have to go on there and manipulate it to get much social benefit. But Facebook has enough information about you to build a model of your social life. If you're active on the site, it knows your friends, family, likes and dislikes, activities, things you enjoy talking about. That's enough to build a predictive capability about you. Facebook could, for instance, be able to predict whether you'd like someone, much the way a good friend who has chunked your personality can automatically know if you'd like a particular person. Take that a step further, and Facebook should be able to suggest matches to single people—not based on a survey, like eHarmony, but based on who you are and the kinds of people you like. Two-second-advantage technology could vault social media to an entirely new level.

We're looking forward to social predictive models blending with mobile technology, particularly GPS. While there's much to be said for doing things online, there's no substitute for seeing a good friend in the flesh. A mobile device could track your location as you were driving around in a car or walking through a big convention or sitting in a restaurant, and it could then let your social model know what you were up to. Presumably, your friends would do the same with their phones and social models. Then your social model could let you know that a friend you tend to chat with about music was in the Guitar Center store you were about to drive past. Sure, there are friend-tracking services like Foursquare out there, but this is an active version—like a friend who knows you. It would make a prediction that you'd like to see this person and engineer real-life serendipity.

It might be easy to think that two-second-advantage technology will benefit only the top tiers of global society—people who have the luxury of doing knowledge work, are able to get good health care, and visit with their pals in high-end guitar shops. But in fact it is changing and will change life in most socioeconomic sectors in most parts of the world, often delivered via an inexpensive mobile phone. One example is a program administered by microinsurer Kilimo Salama to farmers in remote parts of Kenya. These are areas where a good harvest is essential to life and banking is virtually nonexistent. That sets up a potential problem for farmers. If a drought or flood destroys crops, the twentieth-century way for an insurance company to handle it is to wait for the disaster to happen, then issue checks—which the farmers find difficult if not impossible to cash. That would leave the farmers without crops or money. Farmers in turn would hedge their risk each year by saving some money and planting fewer crops, which would guarantee less success in good years.

The microinsurance program takes advantage of a "mobile money" joint venture between Vodafone and Kenyan mobile operator Safaricom. Called M-PESA, the venture is a system for securely sending electronic money to mobile phones, and that money can in turn be used to pay bills, buy groceries, and more. At the very least, M-PESA would be a way to eliminate the problem of paper checks and get money to farmers right after their crops fail. But Kilimo Salama is going a step further and using rainfall reports and other data to predict when weather conditions are about to kill crops and actually send farmers insurance payments before the crops die. The farmers are no longer left with even a moment of panic, wondering whether they'll survive, and then a remarkable thing happens: the farmers make bigger investments in their crops, planting more and keeping less cash in reserve. That means better harvests in good years, which helps boost economic conditions across the region. Predictive technology will increasingly drive targeted ways to make big improvements in billions of lives in the developing world.

Two-second-advantage technology will make a difference in some very big ways. It will make companies operate better, make cities safer and more livable, help the economy run more smoothly, and save lives by helping stop terrorism and disease. It will also make daily personal life better, helping us be more productive, enjoy our free time, and connect with other people who matter to us. The technology is only now emerging, and it will

The Two-Second Advantage and a Better Brain

In the mid-1970s, about the time a preteen Wayne Gretzky was first getting noticed, Canada faced a national identity crisis. The Soviet Union was churning out world-class hockey players and putting together teams that could beat the best Canadian teams. Hockey was and is Canada's national sport—its cultural connective tissue. How could another nation challenge Canada on the world stage? Canada had to find out. It sent university researchers to Soviet sports camps to learn what they did. They discovered that the USSR had built a nationwide system designed to find athletically gifted youngsters and manufacture them into top-tier talent. Much of the focus was on developing superhuman bodies. Reporters typed out endless newspaper columns examining the situation. In 1974 Doug Gilbert of the *Montreal Gazette* visited Moscow and described a kind of electroshock muscle treatment that Soviet sports scientists administered to Soviet hockey players.

"Afterward, the hockey player will feel a warm, perhaps slightly-burning sensation in the area," Gilbert wrote. "But, after twenty sessions conducted over a forty-day period he will, according to Soviet sports medicine, have increased the muscle strength and velocity properties so much that the general performance of the muscle will have increased 50 percent over what could have been done through regular exercise."[149]

Three decades later, when former Soviet sports scientist Sergei Beliaev moved to the United States and started a coaching business, he was still a believer in scientific bodybuilding. "If you think about how the coach is perceived in America, it is primarily psychological," Beliaev said. "In Russia and Europe it is viewed differently. The coach is seen as the human engineer of this person, and has the responsibility to make him able to perform at his best level."[150]

Yet with all the science thrown at developing sports superstars, the Soviet Union never quite produced a Gretzky. The system certainly turned out a lot of great hockey players, but none with the predictive talent and artistry to match Gretzky or his successor in hockey wizardry, Canadian-born Mario Lemieux. The Soviets wanted to engineer the world's greatest hockey players, and they focused on the body. Left to his own devices, Wayne Gretzky—with help from his father, Walter—focused on the mind. It was the mind that won.

In the 2010s young Moscow-born star Alex Ovechkin is becoming perhaps the best Russian hockey player ever. He came of age after the old Soviet sports machine collapsed. In interviews, the free-spirited Ovechkin has made a point of putting down the old science-based system.[151]

So is it possible to engineer a Gretzky? The question pertains not just to hockey but to any field. Have researchers learned enough from recent neuroscience and its intersecting work with computer science to know how to build a star?

If you are raising a child, can science tell you what to do to help him or her become a supertalent at something?

The answer: sort of.

Predictiveness is the essence of talent. Neuroscientists are exploring that more deeply now, but they know that practice is the key to developing predictiveness. That practice has to be *deliberate practice* or *deliberate performance*, systematically strengthening the mental chunks associated with the skill through repetition while also incrementally adding new information and experiences. For a physical talent like hockey or violin playing, the deliberate practice or performance also hones muscle memory that lets the body carry out what the mind predicts.

Scientists claim a typical individual has to practice for about ten thousand hours over a period of years to acquire the predictiveness that makes someone an "expert" in a certain field. As we've seen, the very best—the Gretzkys of any field—hone their predictive powers to the point where they have a metaphorical two-second advantage over anyone else.

The stories of Boston mayor Menino and pickup artist Mystery suggest that talent and the two-second advantage can be developed. Neither man had any apparent special talent or gift in his field when he was younger. Even Gretzky was not an obvious talent as a youth—he was a scrawny boy playing a sport more suited to the strong and swift. All three, over time, constructed their talent and their two-second-advantage capabilities. That they could do it suggests that any of us could do the same.

Any decent pee-wee-league coach or grade-school math teacher knows a version of this already. Practice a lot and you'll get good at something. Practice tirelessly with a laser focus, and you'll get great at something. But there are nuances to this and new ideas that come from knowing what practice and talent look like inside the brain.

Experiences fire neurons and encode information in them. Repeated firings in response to the same sets of information or patterns wire those neurons together. More repetition builds up the connections between neurons so the information moves even faster between them. As that happens, still more repetition fuses patterns into chunks that can access a whole collection of information instantly. As experiences expand and bring in new information, that information gets added to the chunks, eventually creating a complex, sophisticated mental model that assesses a situation in a flash, without having to access all the details stored deep in memory. That's what we call a gut instinct.

At every step, the brain predicts. Accurate predictions make humans feel secure and confident, so then we seek to make our predictions more accurate and more sophisticated. When those more complex predictions prove accurate, we want to ratchet up the level of sophistication. At each step, we get better and faster at predicting what's about to happen, gaining a two-second advantage.

The cycle of chunking, reinforcing, and predicting causes physical changes in the wiring of the brain—a characteristic called neuroplasticity. A brain's software actually changes the brain's hardware, until the brain becomes a finely honed and integrated prediction machine in the person's chosen profession or skill.

But here's the catch—a chicken-and-egg kind of dilemma. Exceptional talent self-selects. Highly talented people are almost always motivated by the fact that they really like what they do. And they like what they do because sometime early in life they found that their predictions in that capability were pretty accurate. A little accuracy in the beginning feels satisfying, and we want more of it. Often, that accuracy means we do something well, and that brings praise or rewards, providing further motivation to seek further experiences and build better predictions. This is why people gravitate toward things they do well—and away from things they don't. This is what people are describing when they say things like "He just took to the sport" or "She seemed like she was born doing that."

But why are those early, nascent predictions accurate in some people and not others? How does that initial spark of interest happen? It's a mystery science can't yet unravel. Something made a young Eduard Schmieder first ask to play the violin when there was no particular musical encouragement from his parents, and something allowed him to have enough success from the start to convince him to begin the cycle of experiences, chunking, predictions, and satisfaction. Wayne Gretzky grew up in Ontario with hockey all around him, so it was not surprising that he tried the sport at a young age. But other kids took to hockey at the same age. If Gretzky had found he was not as good as other kids, he probably would've gotten discouraged and taken up another sport. But something early on allowed him to be good enough and predictive enough to seek more hockey experiences, igniting the spark that turned him into a hockey superstar. Gretzky had the right kind of brain hardware, the right physical attributes, and the right circumstances to help him build a superb hockey mind. While science can explain a lot about how he built his talent after that first spark, it can't really explain how that first spark happened.

Knowing this, what would science tell new parents to do if they wanted to develop a child with a special talent? Drawing on her research on the brains of children, Rutgers neuroscientist Paula Tallal came to a few conclusions that she shared with us:

- You should expose your child to varied experiences so he or she can find that thing that ignites a predictive cycle. Yet don't push for further experiences at the expense of letting your child stay with something he or she likes. That act of dwelling on a skill or experience helps build predictiveness—the first steps toward talent.
- In Tallal's research, language processing is an accurate predictor of future success in school and, later, in almost any intellect-based talent (as opposed to sports or other more physical talents). To ensure the development of language skills in a child, talk to your child a lot, and talk positively. Complex chunking and being predictive with spoken language leads to chunking and predictiveness in reading, which leads to success in academics and most fields. For that reason, Tallal told us, schools should do the opposite of what they usually do: they should encourage kids to talk in class.
- Repetition builds predictiveness, and good predictions generate positive feelings that provide motivation to build yet more predictiveness. So if your child wants to read the same book over and over or play the same

piano piece incessantly or practice the same sports skill—it may drive you as a parent crazy, but that repetition is an important part of developing talent.

- Once a child gravitates toward something—whether it be math, cooking, basketball, piano, writing, or any other skill—give him or her as much opportunity for repetition, deliberate practice, and deliberate performance as possible. After all, ten thousand hours is a lot of hours. Very few people are born with savantlike talent. Most of the world's truly talented people got that way through determined focus and intense practice over a long period of time. One thing brain science hasn't discovered is a shortcut to programming talent into a brain—or, for that matter, into a computer.

In January 1996, Tallal and her Rutgers team, along with brain researcher Michael Merzenich and his team from the University of California at San Francisco, published the findings of their research on children's language skills in the journal *Science*. Their findings were surprising and controversial at the time. They reported that through repetitive drills using personal computers, they could alter children's brains so that children who had difficulty with language could distinguish phonemes and map them correctly to written words. In other words, Tallal's group could create predictiveness and a certain level of talent in language skills in kids who had previously lagged behind their peers. The computer program was rewiring kids' brains. The story was covered by the media worldwide, and in the ten days after the story broke, a reported seventeen thousand calls were made to the Rutgers main number. Almost all the calls were from parents asking to get ahold of the computer program that could fix their children's reading problems.[152]

Picking up on the intense public interest, venture capital investors from E.M. Warburg, Pincus & Co. approached Tallal and her colleagues about turning their scientific discovery into a for-profit business. Together, Tallal's team and Warburg raised enough money to start Scientific Learning Corporation, the first company built around neuroplasticity. The company branded the reading program Fast ForWord, and by 1999 Scientific Learning had had enough success to pull in nearly forty million dollars in an initial public offering. By the end of 2010, Fast ForWord was being offered in six thousand schools in the United States and in schools in forty-four countries. Scientific Learning proved that manufacturing talent in kids could be a good business.

While Tallal focused on children, Merzenich branched off into another difficult question: Can the adult brain be similarly rewired? There had long been consensus that children's brains were malleable and could learn new capabilities. But the popular view was that once a brain was wired, it was pretty much wired for good. The old phrase "You can't teach an old dog new tricks" seemed palpably true.

Merzenich, though, showed that it's not. He zeroed in on reversing cognitive decline in old age by rebuilding the brain's predictiveness and processing speed. In 2003 he formed a company called Posit Science, which created the field of "brain fitness." Like the children who used Scientific Learning's software, older adults engaged with Posit Science's software to play auditory and visual games designed to rewire their brains. A 2009 study, funded by Posit Science and conducted by the Mayo Clinic and the University of Southern California, found that older people (around sixty-five or older) who trained on the company's Brain Fitness Program were twice as fast in processing information, with an average improvement in response time of 131 percent. "What this means is that cognitive decline is no longer an inevitable part of aging," Elizabeth Zelinski of the UC Davis School of Gerontology reported at the time. "Doing properly designed cognitive activities can enhance our abilities as we age."[153]

Posit Science and other popular brain fitness programs—like those available on Nintendo's Wii game console—are intended to reverse a brain's decline, not necessarily build talent anew. Still, the point is that these programs are proving that the brain can be rewired at any age. It should be possible, then, to build the chunks and predictiveness needed to develop talent no matter when you start.

Mayor Menino as an adult assembled an incredible talent for running the city of Boston. Clearly his brain was not wired to be a politician from the time he was a toddler. Pickup artist Mystery was a geeky young man who could barely talk to women. Brick by mental brick, he rewired his brain as an adult and developed a talent for engaging women. It's not hard to come up with other examples of late-blooming talent. Anna "Grandma" Moses famously began painting at age seventy-six and lived to be 101. In the intervening years, she became a celebrated American artist.[154]

Cognitive researcher Gary Klein developed his concept of deliberate performance specifically for adults who have to develop talent while doing their jobs. They often can't afford to spend years going back to school or practicing a new skill. "A loan officer can build her tacit knowledge and intuitive decision-making by exercising them (on the job)," Klein and colleague

Peter Fadde wrote. "She can systematically predict which of the loan applications she forwards will be approved. When her prediction is off—in either direction—she attempts to explain why." Using exercises like this, Klein and Fadde explained, she is "building her mental model and improving what in sports would be called 'next level' skills."[155] By deliberately working to build better chunks and make better predictions, a midcareer worker can become more talented. As scientists research predictiveness and neuroplasticity, they'll likely continue to find more effective ways to build new talent or rejuvenate skills and abilities.

Here's where the computer scientists come in. Whether we're talking about DARPA's SyNAPSE programs, the Blue Brain project in Switzerland, IBM's *Jeopardy!*-playing computer, Rajesh Rao's robots, or talent systems being developed for companies, as researchers try to make computers work more like brains, they have to learn how brains work from the bottom up. That's already adding to our knowledge about brain wiring and brain software. Down the road, Blue Brain and some of the SyNAPSE projects aim to simulate a human brain in a computer, much the way computers now simulate complex systems such as weather. Once a rich brain simulation exists, scientists will be able to test theories far more easily than they can today. A researcher such as Paula Tallal has to put children through months of controlled routines to eventually find out how to alter brain wiring. A simulation would let a researcher write software code, plug it into the computer, and see what happens. That should speed along our understanding of predictiveness and talent and help develop tools and methods to improve our brains.

This is liberating news. No one is stuck with the skills or abilities or brain they have. It's possible to reconstruct a brain to overcome a physical disability or go in a new direction. It's possible at any age to build a two-second advantage. It is possible for us to learn how to develop talent and more capable brains.

In the movie *The Terminator*, just after Sarah Connor finds the location of Cyber Dynamics and learns from Kyle about the company's role in creating machines that eventually go to war with humans, she becomes excited about altering history by "uninventing," as she says, the technology.

"We'll blow up the place . . . burn it down. Something," Sarah says to Kyle.

"Tactically dangerous. We lay low," Kyle retorts.[156]

"Think it through," Sarah argues. "We can prevent the war. Nobody else is gonna do it."

We stand at the brink of a different solution to *Terminator* scenarios. The science-fiction writers have mostly assumed that the human brain will stay as it is while electronic brains catch up and surpass us. Recent science suggests that we can use the advancing machines to become smarter ourselves. While we invent machines that can anticipate the future just a bit, we humans can continue to be like Gretzky, staying two seconds ahead of the competition.

Notes

Wayne Gretzky's Brain in a Box

1. Wayne Gretzky with Rick Reilly, *Gretzky: An Autobiography* (HarperCollins, 1990), New York, N.Y. pp. 83–97.
2. *Playboy* Magazine, *The Playboy Interviews: They Played the Game* (M Press, 2006), New York, N.Y., pp. 175–200.
3. Ibid.
4. *Wayne Gretzky: The Making of The Great One* (Becket Publications, 1998), Dallas, Texas.
5. Malcolm Gladwell, *Blink: The Power of Thinking Without Thinking* (Back Bay Books, 2010), New York, N.Y., p. 10.
6. *Time*, "The Brain Builders," March 28, 1955
7. Stephen Grossberg, interview with Kevin Maney, April 2010.

ONES, TWOS, AND CORTEXES

8. Marc Andreessen and Ben Horowitz, interview with Kevin Maney, originally for a *Fortune* cover story published July 6, 2009.
9. Ben Horowitz, interview with Kevin Maney, spring 2010.
10. Ben Horowitz, ben's blog, http://bhorowitz.com/2010/05/30/how-andreessen-horowitz-evaluates-ceos/.
11. Jeff Hawkins, "Jeff Hawkins on How Brain Science Will Change Computing," TED, February 2003. Available at http://www.ted.com/talks/jeff_hawkins_on_how_brain_science_will_change_computing.html.
12. Ibid.
13. Jeff Hawkins, *On Intelligence* (Times Books, 2004), New York, N.Y., p. 89.
14. Vernon Mountcastle, "The Columnar Architecture of the Neocortex," *Brain*, 1997, p. 701.
15. This explanation is paraphrased from Hawkins, *On Intelligence.*
16. Stephen Grossberg, "The Brain's Cognitive Dynamics: The Link Between Learning, Attention, Recognition, and Consciousness," 2002. Available at http://www.cns.bu.edu/Profiles/Grossberg/MBI/1.html.
17. D. H. Ingvar, "Memory of the Future: An Essay on the Temporal Organization of Conscious Awareness," *Human Neurobiology* 4 (1985): 127–36.
18. Joaquin Fuster, interview with Jeff Garigliano for the authors, April 21, 2010. Fuster is professor emeritus of psychiatry and neuroscience at the UCLA School of Medicine.
19. Stephen Grossberg, interview with the authors, April 23, 2010.
20. Duke University Office of News & Communications, "Brain's Visual Circuits Do Error Correction on the Fly," December 7, 2010. Available at http://news.duke.edu/2010/12/egner_vision.html.
21. Fuster, interview.
22. Grossberg, interview.
23. Paula Tallal, interview with Kevin Maney, January 2011.
24. Children of the Code, http://www.childrenofthecode.org/interviews/tallal.htm.

25. Tallal, interview.
26. Tallal, interview.
27. Beatriz Calvo-Merino et al., "Action Observation and Acquired Motor Skills: An fMRI Study with Expert Dancers," *Cerebral Cortex* 15 (August 2005): 1243–49.
28. University of California, Swartz Center for Computational Neuroscience, "Mobile Brain/Body Imaging (MoBI) of Active Cognition," December 23, 2010.
29. William Duggan, "Strategic Intuition: The Key to Innovation," Columbia Business School, February 6, 2009. Available at: www.portfolio.com/resources/insight-center/2009/02/06/Strategic-Intuition/
30. Ibid.
31. Gary Klein, interview with Kevin Maney, August 2010.
32. Gary Klein, David Snowden, and Chew Lock Pin, unpublished manuscript on "anticipatory thinking," given to Kevin Maney by Gary Klein in August 2010.
33. Ibid.
34. Gladwell, *Blink*, p. 16.
35. Elizabeth Falkner, interviews with Dan Fost for the authors, May and June 2010.
36. Traci des Jardins and Gabriel Maltos, interviews with Dan Fost for the authors, May and June, 2010.

THE TALENTED BRAIN

37. Eduard Schmieder, interview with Kevin Maney, 2010.
38. *Stephen Wiltshire - Prodigious Drawing Ability and Visual Memory,* by Darold Treffert. Available at: http://www.wisconsinmedicalsociety.org/savant_syndrome/savant_profiles/stephen_wiltshire.
39. *Neuropsychological Studies of Savant Skills: Can They Inform the Neuroscience of Giftedness?*, Roeper Review, by Gregory L. Wallace, The Roeper School, 2008.
40. Rita Carter, "Turn off, tune in," *New Scientist*, October 9, 1999.
41. Lawrence Osborne, "Savant for a Day," *New York Times*, June 22, 2003.

42. *The Joy Behar Show*, May 25, 2010. Transcript available at http://archives.cnn.com/TRANSCRIPTS/1005/25/joy.01.html.
43. Mo Rocca, interview with Kevin Maney, July 2010.
44. Jeff Hawkins, interview with Kevin Maney, December 2009.
45. Jim Olds, interview with Kevin Maney, April 2010.
46. Earle Whitmore, interview with Kristin Young for the authors, August 2010.
47. Roger Craig, interview with Kevin Maney, July 2010.

TALENTED SOFTWARE OF THE AVERAGE BRAIN

48. Thomas Menino, interview with Kevin Maney, June 2010.
49. Peggie Gannon, interview with Kevin Maney, June 2010.
50. Abby Goodnough, "Boston's Mayor Faces Foes but Is Still a Favorite," *New York Times*, September 21, 2009.
51. Shelley Gare, "Success Is All in the Mind," *The Australian*, January 24, 2009.
52. K. Anders Ericsson, Roy W. Roring and Kiruthiga Nandagopal, "Giftedness and Evidence for Reproducibly Superior Performance: An Account Based on the Expert Performance Framework," *High Ability Studies*, June 2007.
53. Peter J. Fadde and Gary A. Klein, "Deliberate Performance: Accelerating Expertise in Natural Settings," *Performance Improvement*, October 2010.
54. Tracy Clark-Flory, "The Artful Seducer," *Salon*, August 6, 2007.
55. Mystery with Chris Odom, *The Mystery Method: How to Get Beautiful Women into Bed* (St. Martin's Press, 2007) New York, N.Y. p. 7.
56. Ibid., p. xii.
57. Ibid., p. 5.
58. Ibid., p. 77.
59. R. Douglas Fields, "Watching the Brain Learn," *Scientific American*, November 24, 2009; R. Douglas Fields, "Glia: The New Frontier in Brain Science," *Scientific American*, November 4, 2010.
60. Andy Greenberg, "IBM's Cat-Brain Breakthrough," Forbes.com, November 18, 2009.

61. Greg Fish, "IBM Cat Brain Computer Debunked," Discovery News, December 4, 2009.
62. Joe Lovano, interview with Kevin Maney, 2010.

IF IT ONLY HAD A BRAIN

63. Rajesh Rao, interview with Dan Fost for the authors, 2010.
64. Jeff Hawkins, *On Intelligence* (Times Books, 2004), New York, N.Y., p. 89.
65. Stacey Higgenbotham, "Sensor Networks Top Social Networks for Big Data," *Bloomberg BusinessWeek*, September 14, 2010.
66. SmartData Collective, "Big Data, Big Problems," November 10, 2010. Available at http://smartdatacollective.com/avanade/29625/big-data-big-problems.
67. Ibid.
68. "Most Innovative Companies 2010," *Fast Company*, March 2010.
69. Brett Zarda, "Stopwatches? Sensor Technology Puts the Laptop in Lap," New York Times, July 19, 2010.
70. "Data, Data Everywhere," *Economist*, February 27, 2010.
71. Sumit Chowdhury, interview with Kevin Maney, May 2010.
72. John Gideon, interview with David Gilman for Vivek Ranadivé, June 2010.
73. Jane Johnson (of FICO), interview with Jeff Garigliano for the authors, 2010.
74. Andrew Martin, "Sam's Club Personalizes Discounts for Buyers," New York Times, May 30, 2010.

TALENTED TECHNOLOGY AND TALENTED ENTERPRISES

75. Sharon Adarlo, "E. Orange Strategy, Technology Reduces Crime," The Star-Ledger, March 29, 2009; Richard G. Jones, "The Crime Rate Drops, and a City Credits Its Embrace of Surveillance Technology," New York Times, May 29, 2007.
76. Jose Cordero, interview with Kevin Maney, July 2010.
77. Joel Rubin, "Stopping Crime Before It Starts," *Los Angeles Times*, August 21, 2010.

78. Kemel Delic and Umeshwar Dayal, "The Rise of the Intelligent Enterprise," *Virtual Strategist*, spring 2002.
79. David G. Stork, "Scientist on the Set: An Interview with Marvin Minsky," MIT Press, 2008. Available at http://mitpress.mit.edu /e-books/Hal/chap2/two1.html.
80. Thibaut Scholasch and Sébastien Payen, interview with Dan Fost for the authors, April 2010.
81. Bryan Mistele, interview with the authors, February 2010.
82. Mike Campbell, interview with Dan Fost for the authors, September 2010.
83. Gary Conkright, interview with Dan Fost for the authors, September 2010.
84. Joseph Dupree, interview with Dan Fost for the authors, September 2010.
85. Andrew Lawrence, "Nothing Left to Chance," *Information Age*, January 18, 2007. Available at http://www.information-age.com /channels/information-management/features/272256/nothing-left-to-chance.thtml.
86. Karl Taro Greenfeld, "Loveman Plays 'Purely Empirical' Game as Harrah's CEO," Bloomberg, August 6, 2010. Available at http:// www.bloomberg.com/news/print/2010-08-06/loveman-plays-new-purely-empirical-game-as-harrah-s-ceo.html.
87. Vivek Ranadivé, *The Power to Predict* (McGraw-Hill, 2006), New York, N.Y., pp. 50–51.
88. Randy Huston (of Xcel Energy), interview with Kevin Maney, May 2010.
89. SmartGridCity Web site: http://smartgridcity.xcelenergy.com/ learn/technology-overview.asp.
90. PJM officials, interviews with Don Adams for the authors.
91. Guy Peri, interview with Kevin Maney, August 2010.
92. Edouard Odier, interview with Kevin Maney, September 2010.
93. Jan Marshall, interview with Kevin Maney, October 2010.
94. Department of Homeland Security, Office of Information Technology, "USCIS Faces Challenges in Modernizing Information Technology," September 2005.
95. Leslie Hope, interview with Kevin Maney, September 2010.
96. U.S. Citizenship and Immigration Services, "Backlog Elimination." Available at http://www.uscis.gov/portal/site/uscis/menu

item.5af9bb95919f35e66f614176543f6d1a/?vgnextoid=
68564175bc927210VgnVCM100000082ca60aRCRD&vgnext
channel=9a1d9ddf801b3210VgnVCM100000b92ca60aRCRD.

BRAINY ELECTRONICS AND ELECTRONIC BRAINS

97. John von Neumann, *The Computer and the Brain: Second Edition* (Yale University Press, 1958), New Haven, Conn., pp. xxii–xxvii.
98. Ibid., p. 51.
99. Ibid., pp. 50–51.
100. Steven Pinker, *How the Mind Works* (W.W. Norton & Co., 2009), New York, N.Y., p. 83.
101. James Cameron, *The Terminator*. Script available at http://www.imsdb.com/scripts/Terminator.html.
102. Kwabena Boahen, "Neuromorphic Microchips," *Scientific American*, May 2005.
103. Sandra Aamodt and Sam Wang, "Computers vs. Brains," *New York Times*, March 31, 2009. Available at http://opinionator.blogs.nytimes.com/2009/03/31/guest-column-computers-vs-brains/.
104. Gizmodo, "How Large Is a Petabyte?" http://gizmodo.com/5309889/how-large-is-a-petabyte.
105. Kevin Maney, "Every Move You Make Could Be Stored in a PLR," *USA Today*, September 7, 2004.
106. Michael Merzenich, "Michael Merzenich on Re-wiring the Brain," TED, February 2004. Available at http://www.ted.com/talks/michael_merzenich_on_the_elastic_brain.html.
107. Aamodt and Wang, "Computers vs. Brains."
108. Dharmendra Modha, interview with Jeff Garigliano for the authors, August 2010.
109. Sally Adee, "DARPA's SyNAPSE: Seat of Your Pants-on-a-Chip," IEEE Spectrum, November 21, 2008, http://spectrum.ieee.org/tech-talk/semiconductors/devices/darpas_synapse_seat_of_your_pa.
110. Brian Robinson, "DARPA Seeks to Mimic in Silicon the Mammalian Brain," Defense Systems, November 26, 2008, http://www.defensesystems.com/Articles/2008/11/DARPA-seeks-to-mimic-in-silicon-the-mammalian-brain.aspx.
111. Massmiliano Versace (senior research scientist at the Department

of Cognitive and Neural Systems at Boston University), interview with Jeff Garigliano for the authors, July 2010.

112. Ibid.
113. Modha, interview.
114. Massimiliano Versace, "A Brain Made of Memristors," Brain Blogger, December 18, 2010, http://brainblogger.com/2010/12/18/a-brain-made-of-memristors/.
115. Versace, interview.
116. Ethan Bauley, "HP and Hynix—Bringing the Memristor to Market in Next-Generation Memory," Data Central (HP corporate blog), August 31, 2010, http://h30507.www3.hp.com/t5/Data-Central/HP-and-Hynix-Bringing-the-memristor-to-market-in-next-generation/ba-p/82218.
117. Ibid.
118. Boahen, "Neuromorphic Microchips."
119. Douglas Fox, "Brain-Like Chip May Solve Computers' Big Problem: Energy," *Discover*, November 6, 2009. Available at http://discovermagazine.com/2009/oct/06-brain-like-chip-may-solve-computers-big-problem-energy/article_print.
120. Kevin Maney, "Beyond the PC: Atomic QC," *USA Today*. Available at: www.amd1.com/quantum_computers.html.
121. John Markoff, "Quantum Computing Reaches for True Power," *New York Times*, November 8, 2010.
122. D-Wave Web site, http://www.dwavesys.com/.
123. Jeff Hawkins, interview with Kevin Maney, May 2010.
124. Kevin Maney, "Father of Palm Focuses on Making Computers Brainier," *USA Today*, March 29, 2005.
125. Donna Dubinsky, interview with Kevin Maney, 2005.
126. Hawkins, interview.
127. Hawkins, interview.
128. Hawkins, interview.
129. David Ferrucci, interviews with Kevin Maney, 2010.
130. Ferrucci, interviews.
131. Clive Thompson, "What Is IBM's Watson?" *New York Times*, June 14, 2010.
132. Ibid.
133. Vivek Ranadivé, *The Power to Predict* (McGraw-Hill, 2006), pp. 33–35.

134. Felix Salmon and Jon Stokes, "Bull vs. Bear vs. Bot," *Wired*, January 2011.
135. Ibid.
136. Pinker, *How the Mind Works*, p. 333.
137. Henry Markram, "Henry Markram Builds a Brain in a Super computer," TED, July 2009. Available at http://www.ted.com/talks/lang/eng/henry_markram_supercomputing_the_brain_s_secrets.html.
138. Pinker, *How the Mind Works*, pp. 329–30.
139. Ray Kurzweil, *The Singularity Is Near* (Penguin Group, 2005), New York, N.Y., p. 9.

THE TWO-SECOND ADVANTAGE AND A BETTER WORLD

140. Robert F. Bruner and Sean D. Carr, "Lessons from Wall Street's Panic of 1907," NPR, August 28, 2007. Available at http://www.npr.org/templates/story/story.php?storyId=14004846.
141. Federal Reserve Bank of San Francisco, "What Is the Beige Book, and What Role Does It Play in Setting Interest Rates for Monetary Policy?" November 2003, http://www.frbsf.org/education/activities/drecon/answerxml.cfm?selectedurl=/2003/0311.html.
142. Google.org, Flu Trends, "How Does This Work?", http://www.google.org/flutrends/about/how.html.
143. Jeremy Ginsberg, Matthew H. Mohebbi, Rajan S. Patel, Lynnette Brammer, Mark S. Smolinski, and Larry Brilliant, "Detecting Influenza Epidemics Using Search Engine Query Data," *Nature*, February 19, 2009.
144. Marianne Kolbasuk McGee, "Remote Monitoring Boosts Chronic Illness Care," *Information Week*, October 13, 2010.
145. John Conroy, interview with Kevin Maney, January 2011.
146. Kevin Maney, "Future Search Efforts Will Make Google Look Like 8-Tracks," *USA Today*, March 30, 2004.
147. Scott Morrison, "Google CEO Envisions "Serendipity Engine," *Wall Street Journal*, September 29, 2010.
148. Jeff Bell, interview with Kevin Maney, January 2011.

THE TWO-SECOND ADVANTAGE AND A BETTER BRAIN

149. Doug Gilbert, "Young Talent + Coaching + Technology = Success," *Montreal Gazette*, June 17, 1974.
150. Kyle Jorrey, "Soviet Sports Secrets: Declassified," *Pepperdine University Graphic*, January 31, 2011.
151. Charles McGrath, "Alexander Ovechking, the Mad Russian," *New York Times*, April 9, 2010.
152. Jeffrey M. Schwartz MD and Sharon Begley, *The Mind & The Brain: Neuroplasticity and the Power of Mental Force* (Harper Perennial, 2002), pp. 233–35.
153. University of Southern California, "Computer Exercises Improve Memory and Attention, Study Suggests," *Science Daily*, February 11, 2009, http://www.sciencedaily.com/releases/2009/02/090211161932.htm.
154. "Grandma Moses Is Dead at 101; Primitive Artist 'Just Wore Out,'" *New York Times*, December 14, 1961.
155. Peter J. Fadde and Gary Klein, "Deliberate Performance: Accelerating Expertise in Natural Settings," *Performance Improvement* 49, no. 9 (October 2010): 5–14.
156. James Cameron, *The Terminator*. Script available at http://www.imsdb.com/scripts/Terminator.html.

Index

Brainerd Memorial Library
Haddam, CT 06438
3 6439 00068426 4